LOCALISATION 2011
Proceedings of the Satellite Conference of LT 26

LOCALISATION 2011
Proceedings of the Satellite Conference of LT 26

editor

Stefan Kettemann
POSTECH, South Korea & Jacobs University, Germany

NEW JERSEY • LONDON • SINGAPORE • BEIJING • SHANGHAI • HONG KONG • TAIPEI • CHENNAI

Published by

World Scientific Publishing Co. Pte. Ltd.
5 Toh Tuck Link, Singapore 596224
USA office: 27 Warren Street, Suite 401-402, Hackensack, NJ 07601
UK office: 57 Shelton Street, Covent Garden, London WC2H 9HE

British Library Cataloguing-in-Publication Data
A catalogue record for this book is available from the British Library.

Cover image: A density plot of a multifractal state at the 3D Anderson metal-insulator transition, taken from S. Kettemann, E. R. Mucciolo, I. Varga and K. Slevin, *Phys. Rev. B* **85** (2012) 115112.

LOCALISATION 2011
Proceedings of the Satellite Conference of LT 26

Copyright © 2013 by World Scientific Publishing Co. Pte. Ltd.

All rights reserved. This book, or parts thereof, may not be reproduced in any form or by any means, electronic or mechanical, including photocopying, recording or any information storage and retrieval system now known or to be invented, without written permission from the Publisher.

For photocopying of material in this volume, please pay a copying fee through the Copyright Clearance Center, Inc., 222 Rosewood Drive, Danvers, MA 01923, USA. In this case permission to photocopy is not required from the publisher.

ISBN 978-981-4436-85-4

Typeset by Stallion Press
Email: enquiries@stallionpress.com

Printed in Singapore.

PREFACE

These are the proceedings of Localisation 2011, which took place at POSTECH in Pohang, South Korea, on August 4 to 7th 2011, as a satellite meeting of the 26th International Conference on Low Temperature Physics (LT26). The International Conference Localisation 2011 is the eighth of the series started in Braunschweig in 1984 and since then held in Tokyo, London, Eugene, Jasowiecz, Hamburg and Tokyo again. Traditionally, the purpose of the conference is to provide a forum for discussion of the latest progress on localisation phenomena. The main topics included were: Quantum transport in disordered systems (Anderson localisation, effects of interactions on localisation, Anderson-Mott transition, mesoscopics), Superconductor-insulator transition, Quantum Hall effects (fractional and integer), Topological insulator, Graphene, Dynamical localisation, Heavy fermions (Kondo effect, Kondo lattice, effects of disorder), and Many body localisation (spin-glass, Coulomb glass).

The invited talks were suggested by the international advisory board consisting of Elihu Abrahams (UCLA), Boris Altshuler (Columbia Univ.), P. W. Anderson (Princeton), T. Ando (Tokyo), D. Belitz (Eugene), J. T. Chalker (Oxford), Konstantin Efetov (Bochum), Moty Heiblum (Weizmann Institute, Rehovot), Klaus von Klitzing (MPI Stuttgart), Sergey Kravchenko (Boston), Ad Lagendijk (Amsterdam), Patrick Lee (Boston), Hilbert von Loehneysen (Karlsruhe), R. V. Ramakrishnan (Bangalore), David Thouless (Seattle), Peter Woelfle (Karlsruhe) and Fred Zawadowski (Budapest), and were chosen by the Organizing Committee, consisting of Ravin Bhatt (Princeton University), Mahn-Soo Choi (Korea University), Bernard Coqblin (Orsay, Honorary Chair), Tohru Kawarabayashi (Toho University), Stefan Kettemann (POSTECH/Jacobs University), Hu-Jong Lee (POSTECH), Hyun-Woo Lee (POSTECH), Bernhard Kramer (Chair, Jacobs University), Tomi Ohtsuki (Sophia University), Keith Slevin (Osaka University) and Xin Wan (Zhejiang University).

The program with slides of the presentations can be found at http://www.apctp.org/jrg/activities/LOC11/index.php.

We would like to thank APCTP and WCU-AMS for support, and the staff of APCTP, especially Ms. Jihye Jeong and Ms. Hyunjin Bae, for the friendly and efficient running of the conference.

All papers in this special issue were subject to the peer-reviewing process. We wish to express our gratitude to the authors of the papers and to the advisory and organising committees for their valuable ideas and comments and also their assistance in reviewing the papers.

We dedicate this volume to Professor Bernard Coqblin, CNRS Directeur de Recherche, Honorary Professor of the W. Trzebiatowski Institute of Low Temperature and Structure Research, and a Honorary Chairman of the APCTP Conference Localisation 2011. Professor Coqblin passed away on the 29th of May 2012 at Orsay, France. His support and advice were crucial to the success of this conference. We remember him for his great contributions to condensed matter theory, his good humour and friendship.

Bremen, June 2012
Stefan Kettemann

CONTENTS

PREFACE v

Invited Talks

WAVE PROPAGATION AND LOCALIZATION VIA QUASI-NORMAL MODES
AND TRANSMISSION EIGENCHANNELS
    J. Wang, Z. Shi, M. Davy and A. Z. Genack 1

QUANTIZED INTRINSICALLY LOCALIZED MODES: LOCALIZATION
THROUGH INTERACTION
    P. S. Riseborough 12

ASPECTS OF LOCALIZATION ACROSS THE 2D
SUPERCONDUCTOR-INSULATOR TRANSITION
    N. Trivedi, Y. L. Loh, K. Bouadim and M. Randeria 22

THE SPIN GLASS-KONDO COMPETITION IN DISORDERED
CERIUM SYSTEMS
    S. G Magalhaes, F. Zimmer and B. Coqblin 38

TRANSPORT VIA CLASSICAL PERCOLATION AT QUANTUM HALL
PLATEAU TRANSITIONS
    M. Flöser, S. Florens and T. Champel 49

FINITE SIZE SCALING OF THE CHALKER-CODDINGTON MODEL
    K. Slevin and T. Ohtsuki 60

BULK AND EDGE QUASIHOLE TUNNELING AMPLITUDES
IN THE LAUGHLIN STATE
    Z.-X. Hu, K. H. Lee and X. Wan 70

"RARE" FLUCTUATION EFFECTS IN THE ANDERSON MODEL
OF LOCALIZATION
    R. N. Bhatt and S. Johri 79

Contributed Talks

EFFECT OF ELECTRON-ELECTRON INTERACTION NEAR
THE METAL-INSULATOR TRANSITION IN DOPED SEMICONDUCTORS
STUDIED WITHIN THE LOCAL DENSITY APPROXIMATION
    Y. Harashima and K. Slevin 90

CAN DIFFUSION MODEL LOCALIZATION IN OPEN MEDIA?
    C.-S. Tian, S.-K. Cheung and Z.-Q. Zhang 96

LOCAL PSEUDOGAPS AND FREE MAGNETIC MOMENTS
AT THE ANDERSON METAL-INSULATOR TRANSITION: NUMERICAL
SIMULATION USING POWER-LAW BAND RANDOM MATRICES
    I. Varga, S. Kettemann and E. R. Mucciolo 102

FINITE SIZE SCALING OF THE TYPICAL DENSITY OF STATES
OF DISORDERED SYSTEMS WITHIN THE KERNEL POLYNOMIAL METHOD
    D. Jung, G. Czycholl and S. Kettemann 108

CRITICAL EXPONENT FOR THE QUANTUM SPIN HALL TRANSITION IN Z_2 NETWORK MODEL
    K. Kobayashi, T. Ohtsuki and K. Slevin 114

DISORDER INDUCED BCS-BEC CROSSOVER
    A. Khan 120

A COMPARISON OF HARMONIC CONFINEMENT AND DISORDER IN INDUCING LOCALIZATION EFFECTS IN A SUPERCONDUCTOR
    P. Dey, A. Khan, S. Basu and B. Tanatar 127

QUASI TWO-DIMENSIONAL NUCLEON SUPERFLUIDITY UNDER LOCALIZATION WITH PION CONDENSATION
    T. Takatsuka 133

ENHANCEMENT OF GRAPHENE BINDING ENERGY BY Ti 1ML INTERCALATION BETWEEN GRAPHENE AND METAL SURFACES
    T. Kaneko and H. Imamura 139

GENERALIZATION OF CHIRAL SYMMETRY FOR TILTED DIRAC CONES
    T. Kawarabayashi, Y. Hatsugai, T. Morimoto and H. Aoki 145

ELECTRONIC STATES AND LOCAL DENSITY OF STATES NEAR GRAPHENE CORNER EDGE
    Y. Shimomura, Y. Takane and K. Wakabayashi 151

PERFECTLY CONDUCTING CHANNEL AND ITS ROBUSTNESS IN DISORDERED CARBON NANOSTRUCTURES
    Y. Ashitani, K.-I. Imura and Y. Takane 157

DIRECTION DEPENDENCE OF SPIN RELAXATION IN CONFINED TWO-DIMENSIONAL SYSTEMS
    P. Wenk and S. Kettemann 163

ANALYSIS OF QUANTUM CORRECTIONS TO CONDUCTIVITY AND THERMOPOWER IN GRAPHENE — NUMERICAL AND ANALYTICAL APPROACHES
    A. P. Hinz, S. Kettemann and E. R. Mucciolo 170

INDIRECT EXCHANGE INTERACTIONS IN GRAPHENE
    H. Lee, E. R. Mucciolo, G. Bouzerar and S. Kettemann 177

CRITICAL EXPONENTS FOR ANTIFERROMAGNETIC SPIN CHAINS OBTAINED FROM BOSONISATION
    M. Kossow, P. Schupp and S. Kettemann 183

WAVE PROPAGATION AND LOCALIZATION VIA QUASI-NORMAL MODES AND TRANSMISSION EIGENCHANNELS

JING WANG

Department of Physics, Queens College of the City University of New York
65-30 Kissena Boulevard Flushing, New York, 11367, United States of America
jing.wang@qc.cuny.edu

ZHOU SHI

Department of Physics, Queens College of the City University of New York
65-30 Kissena Boulevard, Flushing, New York, 11367, United States of America
zhou.shi@qc.cuny.edu

MATTHIEU DAVY

Department of Physics, Queens College of the City University of New York
65-30 Kissena Boulevard, Flushing, New York, 11367, United States of America
matthieu.davy@gmail.com

AZRIEL Z. GENACK

Department of Physics, Queens College of the City University of New York
65-30 Kissena Boulevard, Flushing, New York, 11367, United States of America
genack@qc.edu

Field transmission coefficients for microwave radiation between arrays of points on the incident and output surfaces of random samples are analyzed to yield the underlying quasi-normal modes and transmission eigenchannels of each realization of the sample. The linewidths, central frequencies, and transmitted speckle patterns associated with each of the modes of the medium are found. Modal speckle patterns are found to be strongly correlated leading to destructive interference between modes. This explains distinctive features of transmission spectra and pulsed transmission. An alternate description of wave transport is obtained from the eigenchannels and eigenvalues of the transmission matrix. The maximum transmission eigenvalue, τ_1 is near unity for diffusive waves even in turbid samples. For localized waves, τ_1 is nearly equal to the dimensionless conductance, which is the sum of all transmission eigenvalues, $g = \Sigma \tau_n$. The spacings between the ensemble averages of successive values of $\ln \tau_n$ are constant and equal to the inverse of the bare conductance in accord with predictions by Dorokhov. The effective number of transmission eigenvalues N_{eff} determines the contrast between the peak and background of radiation focused for maximum peak intensity. The connection between the mode and channel approaches is discussed.

Keywords: Modes; Transmission channels; Anderson localization.

1. Introduction

The statistics of transmission in samples with length greater than the mean free path are the same for all manner of waves including quantum mechanical waves of electrons and atoms and classical waves such as electromagnetic radiation and ultrasound.[1] Many powerful approaches have been taken to describe the broad array of transport phenomena in random systems. We consider here two approaches which were initially developed to explain the scaling of conductance; one based on quasi-normal modes[2,3] and the other on transmission eigenchannels.[4–7] It has not been possible to make direct observations of the modes and channels in electronic systems; however, these have recently been isolated in microwave experiments[8,9] by analyzing field transmission patterns and have shed light on the character of wave transport. Below, we consider the experimental setup and then discuss the mode and channels representations of transmission and their relationship.

2. Experimental Setup

Microwave measurements are carried out in ensembles of random quasi-one-dimensional samples contained in a copper tube.[8,9] The samples are random mixtures of alumina spheres with diameter of 0.95 cm and index of refraction of $n = 3.14$ at a volume fraction of 0.068. A schematic of the sample tube and the source and detector antennas are shown in Fig. 1. Spectra of the field transmission coefficient polarized along the length of a short wire antenna are obtained from the measurement of the in- and out-of-phase components of the field with use of a vector network analyzer. The sum of intensity across the output face gives the total transmission. The incident field is launched by an antenna that can be translated over a square grid of points on the front surface of the sample. The field speckle pattern for each antenna position on the sample input is obtained by translating the detection antenna on a square grid over the output surface. The source and detector antennas may be rotated in the plane of the surface so that propagation for two polarizations of the wave can be measured. An example of an intensity speckle pattern formed in transmission is shown at the output of the sample tube. The tube is rotated and

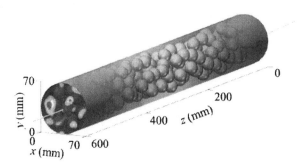

Fig. 1. Copper sample tube filled with scatterers and speckle pattern produced by a wire antenna at the input and detected by a second wire antenna translated over sample output.

vibrated momentarily after measurements are completed on each sample configuration to create a new and stable arrangement of scattering elements. In this way, measurements are made on a random ensemble of statistically equivalent realizations of the sample. Field spectra can be Fourier transformed to yield the temporal response to pulsed excitation.

3. Modes

Excitations in complex media are superpositions of eigenstates that are referred to as 'levels' for quantum systems and 'modes' for classical waves. Anderson showed that electrons in random lattices that are not confined energetically may, nevertheless, be exponentially localized by disorder.[10] Subsequently, Edwards and Thouless argued that the electronic conductance in bounded samples should depend only upon the dimensionless ratio of the average sensitivity of energy levels to changes in the boundary conditions to the average spacing between energy levels, $\delta = \delta E/\Delta E$.[2,3] The Thouless number δ may also be expressed as the ratio of average energy width and spacing of levels or modes of the sample. For classical waves, it is natural to describe the width and spacing between resonances in terms of angular frequency rather than energy, $\omega = E/\hbar$, giving, $\delta = \delta\omega/\Delta\omega$.

The Thouless number reflects the spatial distribution of modes within the sample. When the intensity is then localized within the interior of the sample, the modes are long-lived and the linewidth is small relative to the spacing between modes, $\delta\omega < \Delta\omega$. On the other hand, when modes extend throughout the sample, their amplitudes at the sample boundaries are high so that mode lifetimes are short and the linewidths are broader than the spacing between modes, $\delta\omega > \Delta\omega$. Thus, $\delta = 1$ corresponds to the localization threshold separating diffusive and localized waves.

The decay rate of modal energy Γ_n is the sum of the leakage rate through the sample boundaries, and the absorption rate, $1/\tau_a$, $\Gamma_n = \Gamma_n^0 + 1/\tau_a$ may be determined directly from the linewidths and spacings of modes once the rate of dissipation is determined. The average level width in a sample without dissipation is the inverse of the Thouless time τ_{Th}, and may be expressed in terms of the average of the inverse of the modal leakage rate, $\omega = \tau_{Th}^{-1} = \langle 1/\Gamma_n^0 \rangle^{-1}$.

In the frequency domain, the field at the output surface can be expressed as a superposition of the fields due to the excited modes,

$$E_j(\mathbf{r},\omega) = \sum_n a_{n,j}(\mathbf{r}) \frac{\Gamma_n/2}{\Gamma_n/2 + i(\omega - \omega_n)} = \sum_n a_{n,j}(\mathbf{r})\varphi_n(\omega). \qquad (1)$$

Here, $a_{n,j}$ is the spatial variation of the jth polarization component of the field for the nth mode with central frequency ω_n and linewidth $\Gamma_n\varphi_n(\omega)$ is the frequency variation of the field of the mode, which is proportional to the Fourier transform of for $\exp(-\Gamma_n/2)\cos(\omega_n t)$ $t > 0$, which is the temporal variation of the modal field following delta function excitation at $t = 0$. Since spectra at all $\mathbf{r}$ in a given configuration share a common set of ω_n and Γ_n, the mode expansion in Eq. (1) can be fit simultaneously to spectra at many points to give ω_n, Γ_n, and the corresponding mode speckle patterns.[8]

The number of modes used in the fit of spectra between 10 and 10.24 GHz can be determined unambiguously from the spectra of the squared residual of the fit shown in Fig. 2. Excellent agreement between measurements and the global fit is obtained for both intensity and total transmission. This can be seen in the spectrum of total transmission normalized by its ensemble average for a wave launched by an antenna at point a, $s_a = T_a/\langle T_a \rangle$, shown in Fig. 2(a). Even though the χ^2 goodness of fit improves as the number of modes used in the fit increases and good fits to the total transmission are obtained with different number of modes, the number of modes can be determined unambiguously from the spectrum of the squared residual of the fit (Fig. 2(b)).[11] When the number of modes used in the fit M' matches the actual number of modes $M = M'$, no obvious peak is seen. When an additional mode is then added in the fit, the width of the added mode is unphysically broad with a central frequency that typically falls well beyond the frequency range of the spectrum.

For the spectrum in Fig. 2, $M = 36$, which is larger than the number of transmission peaks that can be made out in this frequency range, indicating that the many modes coalesce to give a single transmission peak. The speckle patterns of

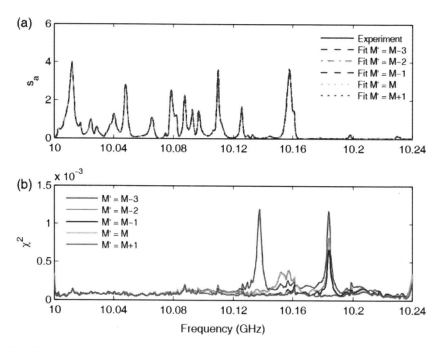

Fig. 2. Modal decomposition. (a) Measured transmission spectrum and fit with different number of modes, M'. (b) Goodness of fit for different M'. The number of peaks falls by one each time a mode is added in the fitting procedure when $M' < M$.

such overlapping modes are correlated to varying degrees. They may be highly correlated as can be seen in the speckle patterns of adjacent modes shown in Fig. 2 of Ref. 7, or more weakly correlated as seen for the two groups of spectrally overlapping modes in Fig. 3. The speckle pattern of mode 21 is seen to bear a similarity to the patterns of each of its two nearest neighbors, while the patterns of modes 23 and 24 are not perceptibly correlated. The degree of correlation is linked to the spatial overlap of resonances from which the modes are formed. This could not be measured in the present quasi-1D tube geometry but could be observed in measurements along the length of single mode waveguides[12].

Figure 3 shows spectra of transmittance over a narrow frequency range in a random sample of length, $L = 40$ cm, as well as of the transmittance of the most prominent modes found from the global fit of Eq. (1) to field spectra. The change in the normalized speckle patterns as the frequency is tuned is shown on the top row of the figure. The mode speckle patterns are shown below the spectra of transmittance. Peaks in total transmission arise when the wave is tuned on or off resonance with modes of the medium. The total transmission on resonance for individual modes can exceed the net total transmission because of the destructive interference between the spectrally overlapping modes which have similar field speckle patterns.[7] The speckle patterns of the first two transmission eigenchannels are shown at the bottom of the figure and will be discussed in section 4 below.

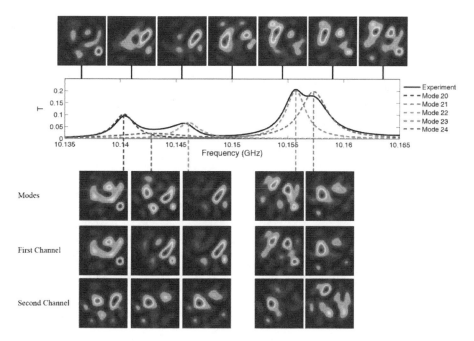

Fig. 3. Spectrum of the transmittance and of the transmittance for the most prominent modes together with intensity speckle patterns for transmission generated by a source at the center of the input surface. Patterns for the modes and first two transmission channels are shown below the spectra. The decomposition into modes and channels is described in the text.

Destructive interference between neighboring modes together with the distribution of mode transmission strengths and widths can explain the total transmission spectrum as well as the time variation of transmission following an incident pulse. The ensemble average of the response of the total transmission to an incident Gaussian pulse is found by composing the response to the pulse from the Fourier transform of the product of the field spectrum and the Gaussian pulse.[13] The progressive suppression of transmission in time by absorption may be removed by multiplying $\langle T_a(t) \rangle$ by $\exp(t/\tau_a)$ to give, $\langle T_a^0(t) \rangle = \langle T_a(t) \rangle \exp(t/\tau_a)$, in which decay is due solely to leakage from the sample. The measured pulsed transmission corrected for absorption in a sample with $\delta = 0.17$, shown as the solid curve in Fig. 4,[8,13] is compared to the incoherent sum of transmission for all modes in the random ensemble corrected for absorption, $\langle \Sigma_n T_{a_n}(t) \rangle$, shown as the dashed curve in the figure[8] $\langle \Sigma_n T_{a_n}(t) \rangle$ is substantially larger than $\langle T_a(t) \rangle$ at early times but then converges to $\langle T_a(t) \rangle$. Though transmission associated with individual modes rises with the incident pulse, transmission at early times is suppressed by destructive interference of modes with correlated field speckle patterns. At later times, random frequency differences between modes lead to additional random phasing between modes and averaged pulsed transmission approaches the incoherent sum of decaying modes. The decay of $\langle T_a(t) \rangle$, shown as the solid curve in Fig. 4, is seen to slow progressively with time delay reflecting the increasing weight of more slowly decaying modes within a broad distribution of decay rates.[8,13]

We have also measured the time variation of intensity correlation on the output surface of the same sample for which measurements of $\langle T_a(t) \rangle$ are shown in Fig. 4. The degree of long-range correlation is seen in Fig. 5 to dip at intermediate times

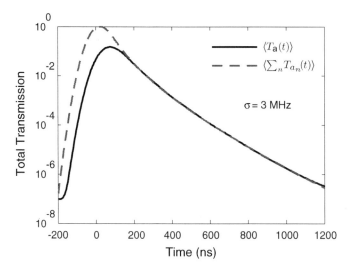

Fig. 4. Ensemble average of time-varying total transmission, $\langle T_a(t) \rangle$, following a Gaussian pulse with linewidth $\sigma = 3$ MHz, and the incoherent sum of transmission due to all modes in the random ensemble. Destructive interference between modes reduces transmission at early times. The curves merge at late times when the phases of the trasnmitted fields due to different modes is randomized.

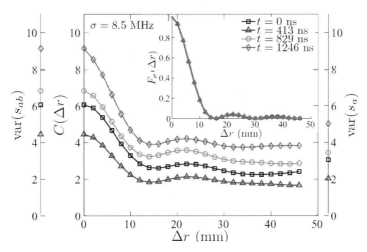

Fig. 5. Variation with time of the cumulant spatial correlation function of intensity, $C(\Delta r)$. The degree of long-range correlation is seen to drop initially and then rise. The value of $C(\Delta r = 0)$ equals the variance of normalized intensity, $\text{var}(s_{ab})$, while the degree of long-range correlation is equal to $\text{var}(s_{ab})$ to within a factor of order $1/N$, where N is the number of transverse propagation channels.

and then to increase.[14] This non-monotonic behavior can be explained in terms of modes. Once the phase between neighboring modes is randomized, the degree of correlation reflects the average of the effective number of modes contributing to transmission. A short time after a small ballistic pulse reaches the output of the sample, the spectrum of transmission is dominated by short-lived modes that release their energy quickly. At long times, short-lived modes have decayed and transmission is dominated by long-lived modes. At an intermediate time, both short and long-lived modes contribute to transmission. Since a larger number of modes contribute to transmission at intermediate times as compared to at earlier or later times, the degree of correlation has a minimum after which the degree of intensity correlation increases. We will see below that a more quantitative description of correlation may be given in terms of the number of independent channels $N_{\textit{eff}}$ because the channels are orthogonal while the transmission speckle patterns for modes are not.

We have recently measured the statistics of mode spacings and widths for microwave radiation localized in samples with δ equal to 0.43 and 0.17 and effective sample lengths of approximately two and three times the localization length, ξ. Even though the wave is localized, the distribution of mode spacings normalized by the average mode spacing, s, is close to the Wigner surmise[15–17] predicted for diffusive waves exhibiting strong level repulsion, $P_W(s) = (\pi s/2)\exp(-\pi s^2/4)$.[11]

4. Channels

The field transmission matrix t relates the incoming and outgoing propagation channels $E_b = \sum_{a=1}^{N} t_{ba} E_a$. These channels may be transverse propagation modes of the empty waveguide or points on a grid over the input and output surfaces. N is the total number of incident transverse modes. The field transmission matrix t can be written as $t = U\Lambda V$, where U and V are unitary matrices and Λ is a diagonal matrix with singular values λ_n of t along the main diagonal. The eigenvalues of the transmission matrix, $tt^\dagger$ are given by $\tau_n = \lambda_n^2$. We find that the ratio between τ_n for adjacent transmission channels is constant.[9] In diffusive samples, most of the eigenvalues are small so that the associated eigenchannels are referred to as "closed", while a number of channels, approximately equal to g, are "open" with transmission eigenvalues greater than $1/e$.[4,5] Thus, it is possible in principle to couple an incident wave to open channels and to achieve nearly full transmission in a multiply-scattering sample in which the average value of the total transmission is small, $\langle T_a \rangle = \langle T \rangle /N = g/N = \langle \tau_i \rangle < 1$. For diffusive waves, we were able to determine up to 25 transmission eigenvalues with the average of the highest of the transmission eigenvalues equal to 0.93.[9] This corresponds to an enhancement for the highest transmission channel over the average value of transmission in microwave experiments of a factor of ~ 10. Spectra of the transmittance and the ten highest transmission eigenvalues in a diffusive sample with $g = 6.9$ are shown in Fig. 6(a). For localized waves, the values of τ_n drop sharply with increasing index n so that the highest transmission channel carries nearly all the transmitted flux giving $\langle \tau_1 \rangle \sim g$. Transmission in the first eigenchannel is then enhanced by a factor of $N \sim 30$.

We find that for both diffusive and localized waves, the spacing between $\langle \ln \tau_n \rangle$ in adjacent channels is constant, $\langle \ln \tau_n \rangle - \langle \ln \tau_{n+1} \rangle \equiv 1/g''$, as can be seen in Fig. 6(b). Dorokhov[4] predicted that g'' is equal to the bare conductance $g'' = g_0 = N\ell/L$. But the bare conductance g_0, which is the value of the conductance neglecting renormalization by weak localization, should be influenced by wave interactions at the sample interface. This may be done by expressing the bare conductance as $g_0 = \eta N\ell/L_{eff}$, where η is a constant of order unity and $L_{eff} = L + 2z_b$ is the effective length, which includes the diffusion extrapolation length z_b at the sample boundaries.[9] The identification of g'' with g_0 is supported by the constant value of

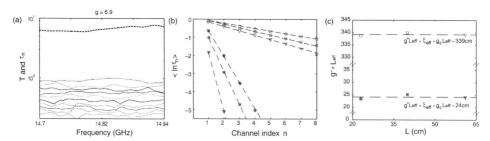

Fig. 6. Spectra of the transmittance T and transmission eigenvalues τ_n for a diffusive sample of length, $L = 23$ cm. (b) Variation of $\langle \ln \tau_n \rangle$ with channel index n for sample lengths $L = 23$ (circle), 40 (square) and 61 (triangle) cm for both diffusive (green open symbols) and localized (red solid symbols) waves, respectively, fitted with black dash lines. (c) The constant products of $g''L_{eff}$ for three different lengths for both diffusive and localized waves gives the localization length ξ in the two frequency ranges.

the products $g''L_{\text{eff}} = \eta N\ell = \xi$, shown in Fig. 6(c) for each of the two frequency ranges studied. This is consistent with a determination of g for the $L = 23$ cm diffusive sample using its relation to the variance of the total transmission normalized by its ensemble average, $\text{var}(s_a - T_a/\langle T_a \rangle)$, which depends upon $P(\tau)$.[18,19] The constant spacing between adjacent values of $\langle \ln \tau_n \rangle$ corresponds to the transmission eigenvalue distribution, $P(\tau) = g/\tau$, peaked as low values of τ in contrast to the bimodal distribution. For the distribution, $P(\tau) = g/\tau$, we obtain, $g = 1/2\text{var}(s_a)$. Measurements of $\text{var}(s_a)$ in this sample, we find $1/2\text{var}(s_a) = 6.6$, which is close to the value $g_0 = \xi/L_{\text{eff}} = 6.9$ expected since $g \to g_0$, in the diffusive limit. The relation between g and $\text{var}(s_a)$ differs from the result found for the bimodal distribution, $g = 2/3/\text{var}(s_a)$.[18,19]

The transmitted speckle pattern for the first transmission channel in resonance with a mode is seen in Fig. 3 to be close to the pattern for the resonant mode. The second channel is necessarily orthogonal to the first and so is made up of speckle patterns of modes that are further off resonance. For this reason the average transmission associated with transmission channels with higher index n is smaller.

The channel representation of transmission also allows for a full description of focusing of radiation through random systems. Using random matrix calculations, the contrast between maximally focused intensity and the background of the transmitted speckle pattern for diffusive waves is found to be, $\mu_N = 1 + N_{\text{eff}}$, where

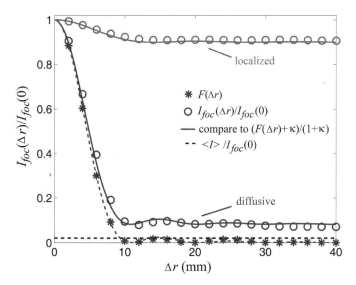

Fig. 7. The ensemble average of normalized intensity for focused radiation (blue circles) is compared to the expression above for the intensity profile of focused radiation (blue solid line) for $L = 61$ cm for diffusive waves. For localized waves for $L = 61$ cm, κ is replaced by $1/(\mu_N - 1)$ in the above expression. $F(\Delta r)$ (blue dots) is fit with the theoretical expression obtained from the Fourier transform of the specific intensity (dashed blue line). The field was recorded along a line with a spacing of 2 mm for 49 input points for $L = 61$ cm. The black dashed line is proportional to $\langle I \rangle / \langle I_{\text{foc}}(0) \rangle = 1/N$.

$N_{\text{eff}} = (\sum_n \tau_n)^2 / \sum_n \tau_n^2$ is the eigenvalue participation number for the transmission matrix.[20] The sum is carried out over all N transverse propagation modes exciting the sample. Maximum focused intensity is achieved by constructing an incident field by phase conjugating the Green's function from the focal point to the incident surface.[21,22] For diffusive waves, N_{eff} is the inverse of the degree of intensity correlation κ. The profile of the focused beam relative to the ensemble average intensity is given by, $\langle I_{\text{foc}}(\Delta r) \rangle / \langle I \rangle = N(F(\Delta r) + \kappa)/1 + \kappa$, where F is the square of the normalized field correlation function. These results are confirmed in microwaves experiments and provide the parameters for optimal focusing and the limits of imaging. Results for diffusive and localized waves are shown in Fig. 7.

5. Conclusion

We have presented an overview of approaches to transport based on modes and channels, which have recently been demonstrated in microwave measurements. Future work will attempt to show that each of these approaches can provide a full picture of propagation. In the mode approach, this will require finding the statistics not only of the spacing between modes and their widths, which have been considered previously,[15–17] but also the degree of correlation between modal speckle patterns and the statistics of modal transmittance. The statistics of transmission eigenchannels have been treated theoretically in previous work,[4–7,18,19,23] but the ability to make direct comparisons of results of random matrix theory calculations with measurements of statistics of individual transmission eigenvalues can test the limits of the applicability of the theory in the Anderson localization transition and support applications such as focusing though random systems. A better understanding of the relationship between the mode and channel approaches will provide a clearer connection between statistics of intensity and total transmission such as is measured in statistical optics and the statistics of more highly averaged quantities such as the conductance. Key additional issues are the spatial distributions of modes and channels and their relation to transmission and reflection and to position-dependent diffusion.[24]

Acknowledgments

We thank Zhao-Qing Zhang and Andrey Chabanov for stimulating discussions. This research was supported by the NSF under Grants No. DMR-0907285 and No. DMR-0958772 MRI-R^2 and by the Direction Générale de l'Armement (DGA).

References

1. A. Lagendijk, B. A. van Tiggelen and D. S. Wiersma, *Physics Today* **62**, 24 (2009).
2. J. T. Edwards and D. J. Thouless, *J. Phys. C: Solid State Phys.* **5**, 807 (1972).
3. D. J. Thouless, *Phys. Rev. Lett.* **39**, 1167 (1977).
4. O. N. Dorokhov, *JETP Lett.* **36**, 318 (1982); *Solid State Comm.* **51**, 381 (1984).

5. Y. Imry, *Euro. Phys. Lett.* **1**, 249 (1986).
6. K. A. Muttalib, J. L. Pichard and A. D. Stone, *Phys. Rev. Lett.* **59**, 2475 (1987).
7. P. A. Mello, P. Pereyra and N. Kumar, *Ann. Phys. (N.Y.)* **181**, 290 (1988).
8. J. Wang and A. Z. Genack, *Nature* **471**, 345 (2011).
9. Z. Shi and A. Z. Genack, *Phys. Rev. Lett.* **108**, 043901 (2012).
10. P. W. Anderson, *Phys. Rev.* **109**, 1492 (1958).
11. J. Wang, Z. Shi and A. Z. Genack, *Phys. Rev. B* **85**, 035105 (2012)
12. P. Sebbah, B. Hu, J. Klosner and A. Z. Genack, *Phys. Rev. Lett.* **96**, 183902 (2006).
13. Z.-Q. Zhang et al., *Phys. Rev. B* **79**, 144203 (2009).
14. J. Wang et al., *Phys. Rev. B* **81**, R241101 (2010).
15. E. P. Wigner, *Proc. Cambridge Phil. Soc.* **47**, 790 (1951).
16. M. L. Mehta, *Random Matrices*, 3rd edn. (Academic Press, New York, 2004).
17. F. J. Dyson, *J. Math. Phys.* **3**, 157 (1962).
18. Th. M. Nieuwenhuizen and M. C. van Rossum, *Phys. Rev. Lett.* **74**, 2674 (1995).
19. E. Kogan and M. Kaveh, *Phys. Rev. B* **52**, R3813 (1995).
20. M. Davy, Z. Shi and A. Z. Genack, *Phys. Rev. B* (2012).
21. I. M. Vellekoop and A. P. Mosk, *Phys. Rev. Lett.* **101**, 120601 (2008).
22. S. M. Popoff et al., *Phys. Rev. Lett.* **104**, 100601 (2010).
23. C. W. J. Beenakker, *Rev. Mod. Phys.* **69**, 731 (1997).
24. C. Tian, S.-K. Cheung and Z.-Q. Zhang, *Phys. Rev. Lett.* **105**, 263905 (2010).

QUANTIZED INTRINSICALLY LOCALIZED MODES: LOCALIZATION THROUGH INTERACTION

PETER S. RISEBOROUGH

Physics Department, Temple University, 1900 North 13th Street
Philadelphia, Pa 19122, USA
prisebor@temple.edu

Received 9 November 2011
Revised 23 February 2012

We have calculated the lowest energy quantized spectra of Intrinsically Localized Modes (ILMs) for the Fermi-Pasta-Ulam lattices. The quantized ILM spectra are composed of resonances in the two-phonon continuum and branches of infinitely long-lived excitations that are bound states formed from even numbers of phonons. For quartic anharmonicity and one atom per unit cell, the calculated ILMs are consistent with the results of previous calculations using the number conserving approximation. However, by contrast the ILM spectrum of the lattice with cubic interactions couples resonantly with the single-phonon spectrum and cannot be calculated within a number conserving approximation. Furthermore we argue that, by introducing a sufficiently strong cubic non-linearity, the quantized ILMs can be observed directly through the single-phonon inelastic neutron scattering spectrum. We compare our theoretical predictions with the recent experimental observation of breathers in NaI by Manley *et al.*

Keywords: Intrinsically Localized Modes; Discrete Breathers; Quantum Mechanics.

PACS numbers: 63.20.Ry, 63.20.Pw

1. Introduction

It is well-known that the quantized elementary excitations of a homogeneous harmonic lattice consist of phonons, which describe vibrations that are extended throughout the entire lattice. It is also known that the presence of impurities can result in localized phonons in which the amplitude of the lattice vibrations are exponentially localized around the impurity sites. It is less well-known that inhomogeneous anharmonic lattices can support persistent vibrations which are localized in space, in the absence of impurities, in which the localization is produced by non-linear interactions. Such excitations are known as either "Intrinsically Localized Modes" (ILMs)[1] or "Discrete Breather Excitations"[2]. Breather excitations are essentially non-linear excitations that were first discovered to occur in integrable continuum field theories such as the Korteweg - de Vries theory[3-4] or the sine-Gordon theory[5-6]. The breather excitations of these integrable theories can be understood as being the bound states of two localized dispersionless waves (solitons) in which the oscillatory motion of the breather is due to the relative motion of the bound soliton and anti-soliton pair. Since the energy of the relative motion is a continuous variable, the excitation energy of the classical ILM has a continuous spectrum (not associated with the continuum spectra due to the center of mass motion), and is bounded from above by the spectrum of free (unbound) soliton / anti-soliton pairs. In addition to the occurrence of breathers in exactly integrable systems, ILMs have also been predicted to occur in many non-integrable classical one-dimensional lattices[7]. Arguments have also been given that classical ILMs should exist in higher-dimensional systems[8], and the existence of ILMs in higher-dimensions are supported by numerical simulations[9]. The existence of ILMs have been inferred either from continuity with the anti-continuous limits of the systems[10] in which the system maps onto an isolated anharmonic oscillator with finite frequency, or from numerical simulations of finite lattices over finite durations. For lattices with one atom per unit cell, the anti-continuous limit requires the existence of a localized potential[11] like that which occurs in the anharmonic Klein-Gordon model, and results in the formation of an optic vibrational mode. However, the simple monoatomic Fermi-Pasta-Ulam (FPU) lattice[12] describes anharmonic acoustic phonons, so it does not posses an anti-continuous limit. An alternate method of proof of the existence of breathers has been given by James[13] who examined time-periodic solutions that can be linearized about the fixed point at which all the oscillators are at rest and, therefore, does not guarantee the existence of large-amplitude breathers. Furthermore, numerical simulations[14] do not guarantee the stability of the large-amplitude localized solutions in either large lattices or over long time scales. On the other hand, as shown by Zabusky and Kruskal[15], the classical FPU lattice maps on to the classical exactly integrable Korteweg - de Vries equation in the continuous limit and, therefore, can be reasonably expected to support large amplitude ILMs.

The theory of quantized ILMs is in a relatively early stage of its development and has been recently reviewed in Ref. 16. Semi-classical quantization of the spectra of ILMs leads to the continuous spectrum associated with the internal oscillations being decomposed into a set of discrete energy levels[17]. These discrete energy levels can be interpreted as either the quantized bound states of a soliton anti-soliton pair, or as bound states of n quantized phonons. For integrable one-dimensional systems, the quantum excitations have been found by use of the quantum inverse scattering method[18-19] which confirms the interpretation of the quantized ILMs in terms of a hierarchy of bound states of n phonons. Since it is expected that the properties of one-dimensional classical field theories more or less carry over to various classical discrete lattice systems which have the same continuum limit[20-21], one might expect that the quantized discrete lattice systems also supports a hierarchy of bound states of the elementary quanta[22]. For systems in which the number of quanta are conserved, the quantized ILM spectra can be obtained analytically[23]. In contrast, for the non-linear Klein-Gordon lattice system for which the phonon number is not conserved, the spectrum has only been obtained by numerical calculation on finite lattices[24-25].

The number of phonons in the FPU lattice system is also not a constant of motion, since the anharmonic interaction can either create or annihilate an even number of phonons. It has been shown[26], by direct construction of an approximate $T = 0$ many-particle wave function, that the quantum FPU lattice supports localized excited vibrational states (i.e. ILMs) which are formed from the bound states of even numbers of phonons. In one-dimension, the branch of bound states first forms for infinitesimal strengths of the anharmonic interaction, at which point, the localization length is infinite and the energy splitting from the upper edge of the two-phonon continuum is infinitesimal. As the strength of the anharmonicity is increased, the localization length rapidly decreases and the energy splitting from the continuum rapidly increases. The ILM dispersion relation gradually flattens with increasing anharmonicity, resulting in a reduction of the group velocity. In this paper, we calculate the two-phonon spectrum using an equation of motion approach involving a T-matrix. The spectrum contains a continuum of two-phonon excitations and branches of resonances and bound states. In the discussion, we argue that the introduction of an appropriately strong cubic anharmonic interaction will result in the quantized ILM spectra being mixed into the spectrum of one-phonon excitations. The results are then compared with recent experimental findings of Manley et al.[29] on three-dimensional NaI crystals.

2. The FPU Hamiltonian

The Hamiltonian for the discrete quartic Fermi-Pasta-Ulam chain can be written as

$$\hat{H} = \sum_i \left[\frac{\hat{P}_i \hat{P}_i^\dagger}{2M} + \frac{M\omega_0^2}{2} \left(\hat{u}_i - \hat{u}_{i+1} \right)^2 \right]$$
$$+ \frac{K_4}{12} \sum_i \left(\hat{u}_i - \hat{u}_{i+1} \right)^4 \qquad (1)$$

in which $\hat{u}_i$ is the displacement operator for the atom at the i-th lattice site from its nominal equilibrium position[a] and $\hat{P}_i$ is the momentum operator for the corresponding atom. The first two terms in the Hamiltonian represent the approximate harmonic Hamiltonian and the third term, proportional to K_4, represents the quartic anharmonic interaction.

The spatial Fourier Transform of coordinates and momenta operators for the atoms are defined as

$$\hat{u}_q = \frac{1}{\sqrt{N}} \sum_i \exp\left[iqR_i \right] \hat{u}_i$$
$$\hat{P}_q = \frac{1}{\sqrt{N}} \sum_i \exp\left[iqR_i \right] \hat{P}_i \qquad (2)$$

---

[a]The ground state is degenerate under the continuous transformation $u_n \rightarrow u_n + \delta$, and therefore the assumed spontaneous symmetry breaking leads to the occurrence of Goldstone modes.

where R_i represents the location of the i-th lattice site. The Hamiltonian can be re-written in the harmonic normal mode basis as

$$\hat{H} = \sum_q \left[\frac{\hat{P}_q \hat{P}_q^\dagger}{2M} + M\omega_0^2 (1 - \cos q) \hat{u}_q \hat{u}_q^\dagger \right]$$
$$+ \frac{K_4}{3N} \sum_{q,k,k'} \left(\cos \frac{q}{2} - \cos k \right) \left(\cos \frac{q}{2} - \cos k' \right) \hat{u}_{\frac{q}{2}+k} \hat{u}_{\frac{q}{2}-k} \hat{u}_{\frac{q}{2}-k'}^\dagger \hat{u}_{\frac{q}{2}+k'}^\dagger \quad (3)$$

It should be noted that the anharmonic interaction has a separable form. The interaction term of the FPU lattice with cubic anharmonicity also is of separable form.

The harmonic part of the Hamiltonian is diagonalized by the transformations

$$\hat{P}_q = i \left(\frac{M \hbar \omega_q}{2} \right)^{\frac{1}{2}} (a_{-q}^\dagger - a_q)$$

$$\hat{u}_q = \left(\frac{\hbar}{2M\omega_q} \right)^{\frac{1}{2}} (a_q^\dagger + a_{-q}) \quad (4)$$

where, respectively, $a_q^\dagger$ and a_q are the boson creation and destruction operators and where the phonon dispersion relation is given by

$$\omega_q^2 = 4\omega_0^2 \sin^2 \frac{q}{2} \quad (5)$$

The dispersion relation describes a branch of Goldstone bosons related to the spontaneously broken translational symmetry. Hence, the Hamiltonian of the FPU lattice can be written as

$$\hat{H} = \sum_q \frac{\hbar \omega_q}{2} (a_q^\dagger a_q + a_q a_q^\dagger) + \hat{H}_{int} \quad (6)$$

where $\hat{H}_{int}$ is the interaction Hamiltonian. The interaction has a separable form and is given by

$$\hat{H}_{int} = \frac{I_4}{6N} \sum_{k_1,k_2,k_3,k_4} \Delta_{k_1+k_2+k_3+k_4} \prod_{j=1}^{4} \left\{ F_{k_j} (a_{k_j}^\dagger + a_{-k_j}) \right\} \quad (7)$$

where the interaction strength I_4 is given by

$$I_4 = \left(\frac{K_4 \hbar^2}{2M^2 \omega_0^2} \right) \quad (8)$$

and the interaction form factor F_k is given by

$$F_k = \frac{\sin \frac{k_j}{2}}{\sqrt{|\sin \frac{k_j}{2}|}} \quad (9)$$

Since the Hamiltonian does not commute with the phonon number operator, the number of phonons is not conserved. Due to the quartic form of the interaction, one expects that the ground state will be a superposition of states with even numbers of phonons.

3. Calculation and Results

We use the equation of motion approach pioneered by Tyablikov and Bonch-Bruevich[27] to obtain the Fourier Transformed matrix of two-phonon time-ordered zero-temperature Green's functions

$$D_{q,k,k'}^{(++)(++)}(t) = -\frac{i}{\hbar} < \hat{T} a_{\frac{q}{2}+k}^\dagger(t) a_{\frac{q}{2}-k}^\dagger(t) a_{\frac{q}{2}-k'}(0) a_{\frac{q}{2}+k'}(0) >$$

$$D_{q,k,k'}^{(+-)(++)}(t) = -\frac{i}{\hbar} < \hat{T} a_{\frac{q}{2}+k}^\dagger(t) a_{-\frac{q}{2}+k}(t) a_{\frac{q}{2}-k'}(0) a_{\frac{q}{2}+k'}(0) >$$

$$D_{q,k,k'}^{(-+)(++)}(t) = -\frac{i}{\hbar} < \hat{T} a_{-\frac{q}{2}-k}(t) a_{\frac{q}{2}-k}^\dagger(t) a_{\frac{q}{2}-k'}(0) a_{\frac{q}{2}+k'}(0) >$$

$$D_{q,k,k'}^{(--)(++)}(t) = -\frac{i}{\hbar} < \hat{T} a_{-\frac{q}{2}-k}(t) a_{-\frac{q}{2}+k}(t) a_{\frac{q}{2}-k'}(0) a_{\frac{q}{2}+k'}(0) > \quad (10)$$

where $\hat{T}$ is the time-ordering operator. Since the interaction has a separable form, the two-phonon Green's functions can be obtained within a standard T-matrix approach without resorting to further approximation[28]. The result can be expressed in terms of the non-interacting Green's functions $D_{q,k,k}^{(\alpha)(0)}(\omega)$, which are diagonal in the pair of indices (α) and (β) and are also diagonal in the Bloch wave vectors k and k' (modulo a reciprocal lattice vector). The result is given by

$$D_{q,k,k'}^{(\alpha)(\beta)}(\omega) = D_{q,k,k}^{(\alpha)(0)}(\omega) \left(\delta^{(\alpha),(\beta)} \Delta_{k-k'} + \delta^{(\tilde\alpha),(\beta)} \Delta_{k+k'} \right)$$

$$+ 2 \left(\frac{\cos\frac{q}{2} - \cos k}{\sqrt{|\cos\frac{q}{2} - \cos k|}} \right) \frac{D_{q,k,k}^{(\alpha)(0)}(\omega) I_4 D_{q,k',k'}^{(\beta)(0)}(\omega)}{1 - I_4 \sum_\gamma \Pi_q^{(\gamma)}(\omega)} \left(\frac{\cos\frac{q}{2} - \cos k'}{\sqrt{|\cos\frac{q}{2} - \cos k'|}} \right)$$

$$(11)$$

where the functions $\Pi_q^{(\beta)}(\omega)$ have been defined as

$$\Pi_q^{(++)}(\omega) = \frac{1}{N} \sum_k \left| \cos\frac{q}{2} - \cos k \right| \left(\frac{1 + N_{\frac{q}{2}+k} + N_{\frac{q}{2}-k}}{\hbar(\omega - \omega_{\frac{q}{2}+k} - \omega_{\frac{q}{2}-k})} \right) \quad (12)$$

and

$$\Pi_q^{(--)}(\omega) = -\frac{1}{N} \sum_k \left| \cos\frac{q}{2} - \cos k \right| \left(\frac{1 + N_{\frac{q}{2}+k} + N_{\frac{q}{2}-k}}{\hbar(\omega + \omega_{\frac{q}{2}+k} + \omega_{\frac{q}{2}-k})} \right) \quad (13)$$

The other two functions are given by

$$\Pi_q^{(+-)}(\omega) = \frac{1}{N} \sum_k \left| \cos\frac{q}{2} - \cos k \right| \left(\frac{N_{\frac{q}{2}-k} - N_{\frac{q}{2}+k}}{\hbar(\omega - \omega_{\frac{q}{2}+k} + \omega_{\frac{q}{2}-k})} \right) \quad (14)$$

and its partner is found from a commutation and the transformation $k \to -k$. The indices (α) and (β) run over the four sets $(\pm, \pm)$ and the tilde denotes transposition.

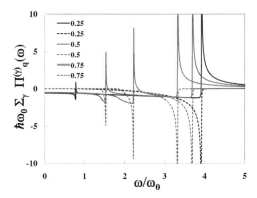

Fig. 1. Plots of the quantities $\lim_{\eta \to 0} \sum_\gamma \Pi_q^{(\gamma)}(\omega + i\eta)$ (in dimensionless units) as a function of dimensionless frequency ω/ω_0, evaluated for various values of q/π (marked in the legend). The real parts are marked by solid lines, and the imaginary parts by broken lines. Due to causality, the van-Hove singularities in the imaginary parts are related to the divergences in the real parts. For a general value of q/π, two van-Hove singularities are present.

In the above expressions, the harmonic phonon frequencies have been set to their temperature-dependent renormalized values[30]

$$\omega_q \approx 2\omega_0 |\sin\frac{q}{2}| \left[1 + \left(\frac{K_4 \hbar}{2M^2 \omega_0^3} \right) \frac{1}{N} \sum_k |\sin\frac{k}{2}| (1 + 2N_k) \right] \quad (15)$$

in which the temperature-dependence is contained in the Bose-Einstein distribution function N_k. The quantity $\lim_{\eta \to 0} \sum_\gamma \pi_q^\gamma(\omega + i\eta)$ is plotted in Fig. 1, and is seen to contain singularities in the real part and imaginary parts. The divergences are truncated due to numerical accuracy. The two-phonon propagator describes collective modes at frequencies ω for which

$$1 - I_4 \sum_\gamma \Re\Pi_q^{(\gamma)}(\omega + i\eta) = 0 \quad (16)$$

where η is a positive infinitesimal number. The collective mode is either an infinitely long-lived ILM if

$$\sum_\gamma \Im m \Pi_q^{(\gamma)}(\omega + i\eta) = 0 \quad (17)$$

or is a broadened resonance if

$$\sum_\gamma \Im m \Pi_q^{(\gamma)}(\omega + i\eta) \neq 0 \quad (18)$$

It is seen that the critical value of I_4, above which ILMs and resonances may exist, is zero. The excitation energies for the ILMs and resonances are determined by the one-dimensional van-Hove singularities in the two-particle density of states, which occur at $\omega = 2\omega_{\frac{q}{2}}$ and at $\omega = 2\omega_{\frac{q+2\pi}{2}}$. The (ω, q) phase-space of two-phonon excitations is shown in Fig. 2. The ILM dispersion relation lies above the top of the two-phonon (creation) continuum. The results found here are in good accord with the results obtained in Ref. 26 using a different method.

4. Discussion

Here we shall argue that an analysis of quantized ILMs similar to that given above might be applicable in three dimensions. At first sight, the existence of classical breathers (when viewed as bound soliton / antisoliton pairs) in three-dimensions seems to be ruled out due to Derrick's theorem[31] which precludes the existence of stable solitons. However, the existence of classical ILMs in three-dimensional systems have not been directly ruled out by Derrick's theorem by virtue of their oscillatory time-dependence. Moreover, the interpretation of quantized ILMs in terms of a hierarchy of bound states is quite robust, so they might very well exist in three-dimensional lattices although some differences may be expected to arise from the dimensionality.

In the above one-dimensional analysis, the quantized ILM occurs for any strength of the quartic anharmonicity, no matter how small, due to the divergence of the van-Hove singularities in $\Pi_q^{(\gamma)}(\omega)$, shown in Fig. 1. Also, the spatial extent of the breathers diverges as the strength of the anharmonicity is reduced[26]. This behavior is similar to the behavior found by Flach, Kladko and MacKay[8] in classical lattices for dimensions less than or equal to two. Their heuristic arguments[8] suggest that ILMs exist in three-dimensional systems for strengths of the anharmonic interaction greater than a critical strength. However, the dependencies of the binding energies and the spatial extents on the interaction strength that are predicted for the three-dimensional ILMs are completely different from the dependencies that the same

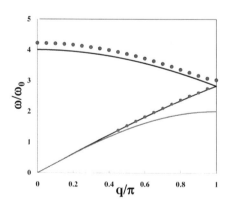

Fig. 2. The (ω, q) phase-space for two-phonon excitations. The two-phonon (creation) continuum is denoted by the blue hatched regions. It is bounded from below the phonon dispersion relation ω_q marked by the red line, and is bounded from above by the van-Hove singularity at $2\omega_{\frac{q+2\pi}{2}}$. The positions of the van-Hove singularities are marked by black lines. The branch of ILMs is denoted by the filled blue markers and the branch of resonances by the filled red markers.

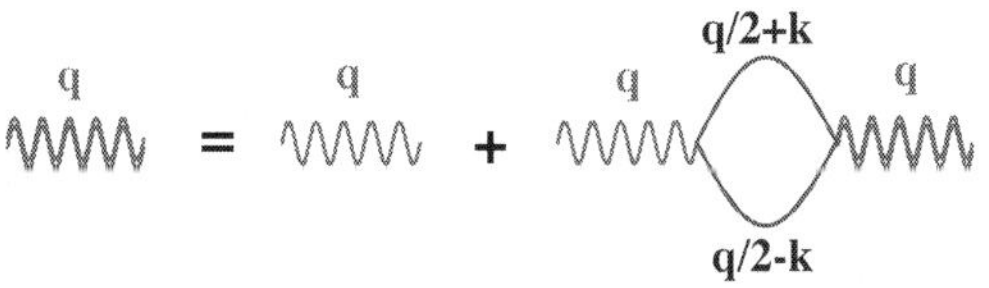

Fig. 3. The simplest Dyson equation relating the fully-renormalized single-phonon propagator of wave vector q (double wavy line) to the single-phonon propagator (single wavy line) and the two-phonon propagator (bubble).

arguments predict for the ILMs of lower-dimensional systems. An extension of the present theory to three-dimensional systems is not straightforward, since the divergences associated with the van-Hove singularities are expected to be washed out. Therefore, in agreement with the arguments made by Flach, Kladko and MacKay, we only expect ILMs to occur in three-dimensions at $T = 0$ if the strength of the anharmonicity I_4 exceeds a large critical value. However, at sufficiently high temperatures, for which the Bose-Einstein distribution function N_q can be replaced by its classical value $(\frac{k_B T}{\hbar \omega_q})$ ($\gg 1$), a generalization of the analysis presented in this manuscript suggests that for values of I_4 below the $T = 0$ critical value, ILMs may still form above a critical temperature determined by an effective anharmonic interaction with a magnitude given by $\sim I_4(\frac{k_B T}{\hbar \omega_0})$. This generalization is consistent with the results of inelastic neutron scattering measurements of Manley et al. on NaI[29]. In the scattering experiments, an additional sharp mode was observed at 10.2 meV above the highest acoustic phonon frequency of 8.8 meV, at $T = 555$ K. However, the mode could not be resolved in measurements performed in the temperature range of 300 to 500 K. This suggests that the strength of the anharmonic interaction is a small fraction of the $T = 0$ critical value. The appearance of the mode as a sharp line, rather than a broad continuum, indicates that if this mode is an ILM, it must be of quantum nature as opposed to being of classical character. For a crystal with purely quartic interactions, the ILM would only be expected to show up as a sharp peak in the two-phonon contribution to inelastic neutron scattering[32-33] and, therefore, would have an intensity proportional to Q^4, where Q is the momentum transfer.

Since NaI does show significant thermal expansion, one infers that odd-order anharmonic interactions such as the cubic anharmonic interaction K_3, must be present. The presence of a cubic interaction would allow for the resonant mixing of the one-phonon and two-phonon spectra, as has been previously noted by Leath and Watson[34]. The occurrence of this type of coupling would then provide a natural explanation of the anomalous softening of the longitudinal acoustic phonon dispersion relation near the zone boundary[29] seen in the experiments of Manley et al.. The lowest-order Feynmann diagram describing this resonant process is depicted in Fig. 3. This resonant coupling would also allow the ILM to appear in the one-phonon contribution to the scattering cross-section. Thus, the integrated intensity of the single-phonon contribution[32–33] to the ILM peak would scale with the momentum transfer as Q^2, as is consistent with the results of Ref. 29.

There have also been reports that ILMs were observed in inelastic scattering experiments on α-uranium[35–38]. Since the anomalous excitations apparently only exist over a small range of total transferred momenta, and since the phonon excitations obey the equipartition theorem[35], the interpretation of the experimental data in terms of localized anharmonic lattice vibrations is a subject of controversy. Therefore, an alternative model for the description of the anomalous vibrational spectrum of α-uranium in terms of the dual character of 5f electrons and a strong electron-phonon interaction has been proposed[39–40]. In that model, the occurrence of the anomalous modes is attributed to a polaronic reduction of the energy scale for electronic excitations[41] to energies comparable to the optic phonon frequencies, at which point, the modes hybridize and produce modes of mixed electronic and phononic characters. The proposed alternate mechanism is not transferrable to NaI. In any case, the data on NaI[29] does appear to be quite consistent with the interpretation in terms of ILMs.

In summary, it seems quite plausible that the mode found by Manley et al. in their high temperature measurements on three-dimensional NaI[29] is in fact a quantized ILM that might be expected from a generalization (to higher dimensions and finite temperatures) of the theory presented here. However, the details of the generalization still remain to be carried out.

Acknowledgments

This work was supported by the U.S. Department of Energy, Office of Basic Energy Sciences, Materials Science through the award DEFG02-84ER45872. The author would also like to thank S. Flach, M.E. Manley and A.J. Sievers for stimulating conversations.

References

1. A.J. Sievers and S. Takeno, *Phys. Rev. Lett.* **61**, 970 (1988).
2. S. Flach and C.R. Willis, *Phys. Rep.* **295**, 181 (1998).
3. D.J. Korteweg and G. de Vries, *Phil. Mag. Series 5*, **39**, 422 (1895).
4. P.S. Riseborough, *Phil. Mag.* **91**, 997 (2011).
5. A. Seeger, H. Donth and A. Kochendörfer, *Zeit. für Phys.* **134**, 173 (1953).
6. J.K. Perring and T.H.R. Skyrme, *Nuclear Physics*, **31**, 550 (1962).
7. S. Flach and A. Gorbach, *Physics Reports*, **467**, 1-116 (2008).
8. S. Flach, K. Kladko and R.S. MacKay, *Phys. Rev. Lett.* **78**, 1207-1210 (1997).
9. S. Flach, K. Kladko and S. Takeno, *Phys. Rev. Lett.*, **79** 4838-4841 (1997).
10. R.S. MacKay and S. Aubry, *Nonlinearity*, **7**, 1623 (1994).
11. J.A. Sepulchre and R.S. MacKay, *Nonlinearity*, **10**, 679 (1997).
12. E. Fermi, J. Pasta and S. Ulam, *"Studies of Nonlinear Problems"*, Unpublished report, Document LA-1940, Los Alamos National Laboratory (May 1955).
13. G. James, *J. Nonlinear Sci.* **13**, 2763 (2003).

14. L.I. Manevitch and V.V. Smirnov, *Phys. Rev. E*, **28**, 036602 (2010).
15. N.J. Zabusky and M.D. Kruskal, *Phys. Rev. Lett.* **15**, 240 (1965).
16. R.A. Pinto and S. Flach, Quantum Discrete Breathers, in *Dynamical Tunneling: Theory and Experiment*, eds. S. Keshavamurthy and P. Schlagheck, (Taylor & Francis, Boca Raton 2011).
17. R.F. Dashen, B. Hasslacher and A. Neveu, *Phys. Rev. D*, **10**, 4114 (1974): *Phys. Rev. D*, **11**, 3424 (1975).
18. L.D. Faddeev and V.E. Korepin, *Phys. Rep.* **42**, 1 (1978).
19. E.K. Skylanin, L.A. Takhtadzhyan and L.D. Faddeev, *Theor. Math. Phys.* **40**, 688 (1979).
20. P.S. Riseborough and S.E. Trullinger, *Phys. Rev. B*, **22**, 4389 (1980).
21. P.S. Riseborough, D.L. Mills and S.E. Trullinger, *J. Phys. C: Sol. St. Phys.* **14**, 1109 (1980).
22. P.S. Riseborough and P. Kumar, *J. Phys. C.M.* **1**, 7439 (1989).
23. A.C. Scott, J.C. Eilbeck and H. Gilhoj, *Physica D*, **78**, 194 (1994).
24. W.Z. Wang, J. Tinka Gammel, A.R. Bishop and M. Salkola, *Phys. Rev. Lett.* **76**, 3598 (1996).
25. L. Proville, *Phys. Rev. B*, **71**, 104306 (2005): L. Proville, *Europhys. Lett.* **69**, 763 (2005).
26. S. Basu and P.S. Riseborough, *Phil. Mag.* **92**, 134-144 (2012).
27. S.V. Tyablikov and V.L. BonchBruevich, *Advances in Phys.* **11**, 317-348 (1962).
28. P.S. Riseborough, *Phys. Rev. E.* **85**, 011129 (2012).
29. M.E. Manley, A.J. Sievers, J.W. Lynn, *et al.*, *Phys. Rev. B*, **79**, 134304 (2009).
30. P.S. Riseborough, *Solid State Commun.* **48**, 901 (1983).
31. G.H. Derrick, *J. Math. Phys.* **5**, 1252 (1964).
32. S.W. Lovesey, *Theory of Neutron Scattering from Condensed Matter*, (Oxford University Press, Oxford 1984).
33. A.A. Maradudin and A.E. Fein, *Phys. Rev.* **128**, 2589 (1962).
34. P.L. Leath and B.P. Watson, *Phys. Rev. B*, **3**, 4404 (1971).
35. M.E. Manley, B. Fultz, R.J. McQueeney, *et al.*, *Phys. Rev. Lett.* **86**, 3076 (2001).
36. M.E. Manley, G.H. Lander, H. Sinn *et al.*, *Phys. Rev. B*, **67** 052302 (2003).
37. M.E. Manley, M. Yethiraj, H. Sinn, *et al.*, *Phys. Rev. Lett.* **96**, 125501 (2006).
38. Michael E. Manley, Jeffrey W. Lynn, Ying Chen and Gerard H. Lander, *Phys. Rev. B*, **77**, 052301 (2008).
39. P.S. Riseborough and X-D. Yang, *J. Mag. Mag. Mat.* **310**, 938 (2007).
40. X-D. Yang and P.S. Riseborough, *Phys. Rev. B*, **82**, 094303 (2010).
41. X-D. Yang, P.S. Riseborough and T. Durakiewicz, *J. Phys. CM*, **23**, 094211 (2011).

ASPECTS OF LOCALIZATION ACROSS THE 2D SUPERCONDUCTOR-INSULATOR TRANSITION

NANDINI TRIVEDI[1], YEN LEE LOH[2], KARIM BOUADIM[3], and MOHIT RANDERIA[1]

[1]*Department of Physics, The Ohio State University, Columbus, OH 43210, USA*
E-mail: trivedi.15@osu.edu
http://www.physics.ohio-state.edu/~trivedi

[2]*Department of Physics and Astrophysics, University of North Dakota Grand Forks, ND 58202, USA*

[3]*Institute for Theoretical Physics III, University of Stuttgart, Germany*

It is well known that the metal-insulator transition in two dimensions for non-interacting fermions takes place at infinitesimal disorder. In contrast, the superconductor-to-insulator transition takes place at a finite critical disorder (on the order of $V_c \sim 2t$), where V is the typical width of the distribution of random site energies and t is the hopping scale. In this article we compare the localization/delocalization properties of one and two particles. Whereas the metal-insulator transition is a consequence of single-particle Anderson localization, the superconductor-insulator transition (SIT) is due to pair localization – or, alternatively, fluctuations of the phase conjugate to pair density. The central question we address is how superconductivity emerges from localized single-particle states. We address this question using inhomogeneous mean field theory and quantum Monte Carlo techniques and make several testable predictions for local spectroscopic probes across the SIT. We show that with increasing disorder, the system forms superconducting blobs on the scale of the coherence length embedded in an insulating matrix. In the superconducting state, the phases on the different blobs are coherent across the system whereas in the insulator long-range phase coherence is disrupted by quantum fluctuations. As a consequence of this emergent granularity, we show that the single-particle energy gap in the density of states survives across the transition, but coherence peaks exist only in the superconductor. A characteristic pseudogap persists above the critical disorder and critical temperature, in contrast to conventional theories. Surprisingly, the insulator has a two-particle gap scale that vanishes at the SIT despite a robust single-particle gap.

Keywords: Metal-insulator transition; superconductor-insulator transition; localization; phase fluctuations.

PACS numbers: 74.25.-q,72.15.Rn,02.70.Ss,64.70.Tg, 05.30.Rt,42.50.Lc

1. Introduction

A superconductor (SC) is an emergent state of matter in which electrons pair up forming Cooper pairs, the different Cooper pairs become phase coherent, and the system undergoes Bose-Einstein condensation. What is the effect of disorder on such a phase-coherent state? It was argued by Anderson[1] that three-dimensional superconductivity is quite robust, persisting even in polycrystalline or amorphous materials. Two dimensions turns out to be particularly intriguing because it is the marginal dimension for localization and superconductivity. One can ask the question the other way around: starting with a two-dimensional disordered system in which all the single-particle states are localized, how does it develop superconductivity when attractive interactions between electrons are turned on? What is the specific mechanism[2] that generates superconductivity in a localized system? It is seen from experiments that superconductivity in two dimensions does exist but can be destroyed by a large variety of tuning parameters including temperature, inverse thickness (characterized by sheet resistance), disorder, gate voltage, Coulomb blockade, perpendicular magnetic field, and parallel magnetic field[3–15]. The destruction of superconductivity is a quantum phase transition[16] occurring at zero temperature. In this article we provide answers to the following questions related to the superconductor-insulator transition (SIT):

(1) For zero disorder, we know that above the superconducting transition temperature T_c the system is a normal Fermi liquid. What is the nature of the state above T_c at finite disorder? Is it a Fermi liquid?
(2) What is the nature of the insulator? Is it a localized Anderson insulator, a Mott insulator, a Fermi glass, a Bose glass, or something else?
(3) Is multifractality of the single particle states important at the SIT?
(4) What are the energy scale(s) in the insulator that vanish at the SIT?
(5) What is the mechanism that drives the SIT?
(6) How do the single-particle spectral functions and dynamical conductivity behave in the superconductor, the insulator, and near the SIT?

We start with a definition of the Anderson model of localization and the nature of non-interacting states at different energies. We then discuss the disordered attractive Hubbard model and the nature of many-particle states obtained by Bogoliubov de Gennes mean field theory[17,18]. We augment the inhomogeneous mean field theory with quantum Monte Carlo simulations and maximum entropy techniques to extract one-particle and two-particle spectral information[19]. We specifically address the role of amplitude variations in a random environment and phase fluctuations of the order parameter in generating superconducting and insulating phases. We conclude with ideas for future experiments.

2. Anderson Model of Localization

Consider a tight-binding model with a disorder potential v_i at each site, chosen independently from a uniform distribution on $[-V, +V]$, where V is the disorder strength. This is known as the "Anderson model" of localization:

$$H = -\sum_{ij} t_{ij} c_i^\dagger c_j + \sum_i (v_i - \mu) n_i. \tag{1}$$

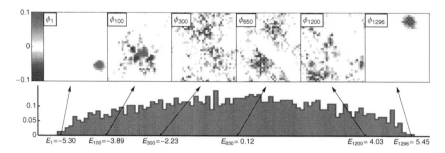

Fig. 1. Six eigenstates of the Anderson model on a 36 × 36 square lattice for a single disorder realization ($\mu = 0, V = 3t$). Red and blue colors indicate signs of eigenfunctions $\psi_{i\alpha}$.

The hopping alone would produce plane-wave eigenstates with a bandwidth of $2zt$, where z is the coordination number, whereas the disorder potential alone would produce site-localized eigenstates with a bandwidth of $2V$. The competition between hopping and disorder makes this a non-trivial problem.

Figure 1 shows selected eigenvectors of the Anderson model for a given realization of the random potentials. The eigenvectors correspond to energies in the tails and in the middle of the density of states. The states in the band tail are localized, whereas the states in the band center appear to be extended over the size of the system; however, in an infinite-size system, they would be localized[20,21].

For a two-dimensional system with on-site disorder, it is well known that an infinitesimal disorder strength, V, is enough to localize all single-particle eigenstates. That is, there is a metal-insulator transition occurring at infinitesimal V.

The localization length ξ_{loc} is the length scale for the exponential decay of an eigenfunction far from its center of mass. It can be obtained from the decay length of the transmission coefficient along a long strip calculated using transfer matrix methods or Green function methods[22]. The localization length ξ_{loc} is finite for any finite V and decreases as V increases (see Fig. 2).

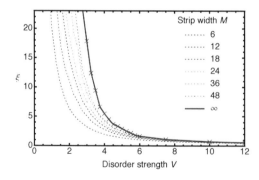

Fig. 2. Localization length of the 2D Anderson model (blue curve), estimated using a crude form of finite-size scaling on Green functions of long strips.

3. Disordered Attractive Hubbard Model

We now turn to the main topic of this paper: a disordered superconductor. We represent this by an attractive Hubbard model with a disorder potential. Alternatively, one may think of this as the Anderson model of localization plus an on-site Hubbard attraction, $|U|$:

$$H = -\sum_{ij\sigma} t_{ij} c_{i\sigma}^\dagger c_{j\sigma} - \sum_i \mu_i n_{i\sigma} - |U| \sum_i c_{i\uparrow}^\dagger c_{i\downarrow}^\dagger c_{i\downarrow} c_{i\uparrow}, \qquad (2)$$

where $\mu_i = v_i - \mu$, where the disorder potential at each site v_i is picked independently from a uniform distribution on the interval $[-V, +V]$, as before.

The attractive interaction $|U|$ has profound consequences. In the absence of disorder, it is well known that an infinitesimal $|U|$ is sufficient to produce Bardeen-Cooper-Schrieffer (BCS) pairing, which leads to the phenomenon of superconductivity. The superconducting state, moreover, exhibits quasi-long-range order up to a finite temperature T_{BKT}, where it is destroyed by a Berezinskii-Kosterlitz-Thouless (BKT) transition involving vortex-antivortex unbinding.

3.1. *Atomic Limit*

In the limit of extreme disorder, the hopping can be neglected, and the system then reduces to an ensemble of single-site Hubbard models, each with the Hamiltonian

$$H = -|U| n_\uparrow n_\downarrow + (V - \mu)(n_\uparrow + n_\downarrow). \qquad (3)$$

This system has just four Fock states. The energies of these states are $E_0 = 0$, $E_\uparrow = E_\downarrow = V - \mu$, and $E_{\uparrow\downarrow} = -|U| + 2(V - \mu)$. The four states occur with relative Boltzmann weights $\exp(-\beta E_n)$. The spectral function (the density of states for single-particle excitations) can be obtained by considering transitions between these four Fock states (amplitudes and energies). This is illustrated in Fig. 3. The ground state is always either doubly occupied or empty. Regardless of the on-site potential V, single-particle transitions (black arrows) always cost at least $|U|/2$, and therefore the spectrum is always gapped. [a] Pair excitations (purple arrows) correspond to transitions from or vice versa. At the specific matching value $\mu_i = v_i - \mu = 0$ these pair excitations may cost zero energy.

We have generalized the above calculation to exact diagonalization of the many-body Hubbard Hamiltonian on small clusters of a few sites, which leads to the same conclusions: single-particle excitations are gapped whereas two-particle excitations can become gapless for an appropriate choice of $\{v_i - \mu\}$.

We know that a clean s-wave superconductor ($V = 0$) has a gap $E_g = \Delta$ given by the BCS gap equation. We have just found that in the limit of extreme disorder, $V \gg (|U|, t)$, the gap is finite and large, $E_g = |U|/2$. We willl see that the gap remains finite between these two extremes.

---

[a]This is in contrast to the repulsive-U Hubbard model in the atomic limit, for which the spectrum is only gapped if $U > 2V$.

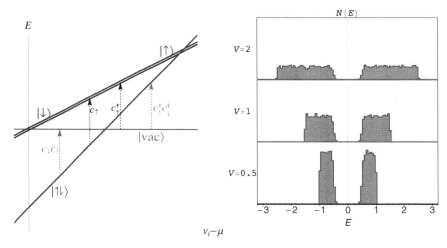

Fig. 3. (Left) Energy levels E and single-particle and two-particle transition energies for a single site with potential v_i, chemical potential μ, and attraction $|U|$ – the atomic limit of the Hubbard model. (Right) Density of states of a 100×100 lattice in the atomic limit ($t = 0$) with attraction $|U| = 1$. At zero temperature the DoS has a hard gap $|U|/2$ regardless of disorder strength V. At finite temperature ($T = 0.05$ in the figure above) there is a small amount of weight in the gap due to thermal excitations.

3.2. *Pairing of Exact Eigenstates (PoEE)*

The above Hamiltonian contains three terms: hopping, disorder, and attraction. In typical *s*-wave superconductors, the first two terms have the largest energy scales. Thus, it makes sense to solve the non-interacting problem first by direct diagonalization, to find the disorder eigenvalues and eigenstates ξ_α and $\phi_{i\alpha}$ and then examine the effect of $|U|$. This is very much in the spirit of the derivation of Anderson's theorem[1].

In the basis of exact eigenstates, the Hamiltonian is

$$H = \sum_\alpha \xi_\alpha \gamma^\dagger_{\alpha\sigma} \gamma_{\alpha\sigma} - |U| \sum_{\alpha\beta\gamma\delta i} \phi_{i\alpha}\phi_{i\beta}\phi^*_{i\gamma}\phi^*_{i\delta} c^\dagger_{\alpha\uparrow} c^\dagger_{\beta\downarrow} c_{\gamma\downarrow} c_{\delta\uparrow}. \qquad (4)$$

Following Anderson's suggestion, let us assume that instead of pairing between k and $-$k, we have pairing between time-reversed eigenstates α and $\bar\alpha$ (i.e., complex conjugate eigenfunctions). Retaining only those terms in the Hamiltonian that connect such eigenstates we obtain the gap equation[17,18]

$$H_{\text{PoEE}} = \sum_\alpha \xi_\alpha \beta^\dagger_{\alpha\sigma}\beta_{\alpha\sigma} - |U|\sum_{\alpha\beta i}\phi_{i\alpha}\phi_{i\bar\alpha}\phi^*_{i\bar\beta}\phi^*_{i\beta} c^\dagger_{\alpha\uparrow}c^\dagger_{\bar\alpha\downarrow}c_{\bar\beta\downarrow}c_{\beta\uparrow}$$

$$= \sum_\alpha \xi_\alpha \beta^\dagger_{\alpha\sigma}\beta_{\alpha\sigma} - \sum_{\alpha\beta} M_{\alpha\beta} c^\dagger_{\alpha\uparrow}c^\dagger_{\bar\alpha\downarrow}c_{\bar\beta\downarrow}c_{\beta\uparrow} \qquad (5)$$

where $M_{\alpha\beta} = |U|\sum_i |\phi_{i\alpha}|^2 |\phi_{i\beta}|^2$. Approximate this by a mean-field Hamiltonian

$$H_{\text{MF}} = \sum_\alpha \xi_\alpha \beta^\dagger_{\alpha\sigma}\beta_{\alpha\sigma} - \sum_\beta (\Delta^*_\beta c_{\bar\beta\downarrow}c_{\beta\uparrow} + h.c.) \qquad (6)$$

(up to a constant), where the order parameter is

$$\Delta^*_\beta = |U| \sum_\alpha M_{\alpha\beta} \langle c^\dagger_{\alpha\uparrow} c^\dagger_{\bar\alpha\downarrow} \rangle \tag{7}$$

(assuming that ξ_α have been redefined in this step to include Hartree shifts).

The gap equation works out to be

$$\Delta_\alpha = |U| \sum_\beta M_{\alpha\beta} \frac{\Delta_\beta}{2E_\beta} \tanh \frac{E_\beta}{2T} \tag{8}$$

where $E_\beta = \sqrt{\xi_\beta{}^2 + \Delta_\beta{}^2}$, and the chemical potential is determined by the number equation

$$\langle n \rangle = \frac{1}{N} \sum_\alpha \left(1 - \frac{\xi_\alpha}{E_\alpha} \right). \tag{9}$$

The PoEE theory can be used in the above form, or one can perform further approximations as follows. In the low-disorder regime, the disorder eigenstates $\phi_{i\alpha}$ are extended on the scale of the system, so that $M_{\alpha\beta} \approx 1/N$ independent of α and β. In this limit Anderson's theorem applies – the gap equation takes the simple BCS form, and Δ is spatially uniform. In the high-disorder regime, on the other hand, the disorder eigenstates are strongly localized with localization lengths ξ^{loc}_α, and the M matrix is approximately diagonal, $M_{\alpha\beta} \approx \delta_{\alpha\beta} \sum_i |\phi_{i\alpha}|^4 \approx \delta_{\alpha\beta}/(\xi^{\text{loc}}_\alpha)^2$.

Surprisingly, *the gap is finite for all values of disorder* $0 < V < \infty$. At large disorder, the gap increases with disorder as

$$E_g = \frac{|U|}{2\xi_{\text{loc}}{}^2}, \tag{10}$$

where ξ_{loc} is the localization length at the chemical potential: as the single-particle states become more localized, the effective attraction is enhanced, leading to a larger gap. [b]

There have been proposals[23,24] that the multifractal nature of the single particle states modifies the exponent in Eq. (10) to the fractal dimension $d_f = 1.7$ in 3D. It is not entirely clear how applicable the fractal nature of the eigenstates is in two dimensions, where the metal-insulator transition which occurs at $V^{\text{MIT}}_c = 0^+$ and the superconductor-insulator transition at $V^{\text{SIT}}_c \approx 2t$ are widely separated. More importantly, it should be remembered that this is a mean-field analysis and there are considerably more important changes to the many-particle states introduced by phase fluctuations.

It has also been proposed[24,25] that T_c gets enhanced near the SIT and can in fact increase without bound. These statements and calculations assume that the SC transition temperature is determined by the gap which as we have discussed is

---

[b]The model Hamiltonian does not include Coulomb repulsion. In real materials it is possible that the Finkel'stein mechanism – disorder-enhanced Coulomb repulsion – may compete with the disorder-enhanced attraction mechanism described here. A full analysis remains to be done.

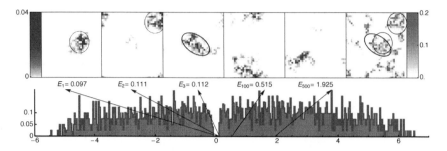

Fig. 4. Eigenvalues and eigenvectors of a disordered superconductor (see text for description of model). The first five panels show the magnitude of five BdG eigenstates (bogolon wavefunctions), $|u_i|^2 + |v_i|^2$. The last panel (red) is a map of the local pairing amplitude Δ_i. The low- and high-energy eigenstates are localized, whereas the intermediate-energy eigenstates are quasi-extended. In particular, the lowest eigenstates correspond to the locations of the superconducting puddles (where Δ_i is large). The parameters are $|U| = 1.5t$, $n = 0.875$, $N = 36 \times 36$, $V = 3t$.

incorrect. Within BCS theory in a weakly coupled clean SC the transition temperature T_c is indeed determined by the gap scale. But that situation changes entirely in a strongly coupled SC or even in a weakly coupled but disordered SC as we have shown. The gap remains finite across the SIT but the phase stiffness goes soft and ultimately vanishes at the transition and it is the phase stiffness scale that now determines T_c. So while a temperature scale associated with the gap may increase near the SIT, the true transition T_c at which the resistance vanishes decreases monotonically with disorder and vanishes at the SIT.

The PoEE approach is useful for understanding the robustness of the gap, but it does not give the full story. It predicts the BCS coherence peaks in the density of states survive up to infinite disorder, whereas more accurate calculations show that there are significant pile-ups in the density of states only in the superconductor. Furthermore, being a mean-field theory, PoEE fails to capture the destruction of phase coherence at SIT due to quantum phase fluctuations. We now proceed to more detailed discussion of phase fluctuations.

4. BdG and QMC results

4.1. *Eigenstates*

Let us consider the eigenvalues and eigenvectors of a dirty superconductor within Bogoliubov-de Gennes (BdG) inhomogeneous mean-field theory, as shown in Fig. 4. The eigenvectors are considerably more localized than for the non-interacting case (Fig. 1). This is clearly seen when comparing Fig. 1 ($U = 0$) and Fig. 4 ($|U| = 1.5t$), which are both at the same disorder strength, $V = 3t$. The eigenvalue spectrum is gapped. How then does this system sustain superconductivity?

We will see that both amplitude and phase fluctuations play an integral role in driving the SIT.

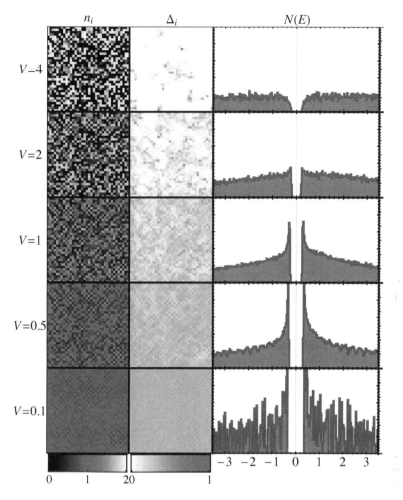

Fig. 5. The three columns show BdG results for the number density, pairing density, and single-particle spectrum of the attractive Hubbard model with a disorder potential, on a 36 × 36 square lattice with $|U| = 2$, $\langle n \rangle = 0.875$, and $T = 0$, as functions of disorder strength V. Here $t = 1$.

4.2. Amplitude vs. Phase Fluctuations

A singlet s-wave superconductor is described by a complex order parameter $\Delta(\mathbf{R}) = |\Delta(\mathbf{R})| e^{i\theta(\mathbf{R})}$. At zero temperature the pairing amplitude $|\Delta(\mathbf{R})|$ takes a uniform value, Δ_0. This is the energy scale associated with pairing. It typically manifests itself as an energy gap $E_g = \Delta_0$, and it also sets the maximum temperature, $T_{\text{pair}} = 0.57\Delta_0$, for the formation of Cooper pairs.

On the other hand, the fluctuations of the phase, $\theta(\mathbf{R})$, are controlled by the superfluid density (or phase stiffness) ρ_s. For 2D superconductors ρ_s has the dimensions of energy, and it can be directly interpreted as the energy scale for phase

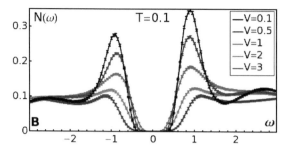

Fig. 6. Disorder dependence of single-particle spectrum from QMC+MEM at very low temperature. With increasing disorder, quantum phase fluctuations eventually wash out the coherence peaks, but the gap is robust.

fluctuations. ρ_s can be measured using mutual inductance techniques. It also sets the maximum temperature, T_{phase}, for long-range phase coherence.

What will emerge from the discussions below is that for a clean superconductor ($V \to 0$), the single-particle eigenstates are "localized" by attraction, on the scale of the superconducting *coherence length* ξ_{coh}^0. At weak disorder the pairing amplitude is homogeneous. At strong disorder the system breaks up into blobs as seen in Fig. 4 and the system can be described by a granular superconductor or a Josephson Junction Array (JJA), where phase fluctuations are extremely important.

What is the size of these blobs? It is given by the coherence length ξ_{coh} defined as the scale over which the order parameter bends and is modified to include the effects of disorder. In the limit of large disorder ($V \to \infty$), the single-particle eigenstates are dominated by Anderson localization, and are localized on the scale of ξ_{loc}. Thus $\xi_{\text{blob}} = \min[\xi_{\text{coh}}, \xi_{\text{loc}}]$. For weak disorder, the phases of the order parameter on different blobs are coupled leading to a globally phase coherent superconducting state. On the other hand at strong disorder the phases on the different blobs lose long range phase coherence and the system becomes an insulator (see Fig. 10). As the quantum phase transition is approached both the correlation length $\xi \approx \xi_{\text{blob}} \delta^\nu$ and the correlation time $\xi_\tau \approx \xi_{\text{blob}} \delta^{z\nu}$ diverge. Here ν and z are critical exponents.

4.3. *Emergent Granularity*

In the Finkel'stein mechanism[26] of the superconductor-insulator transition, the pairing amplitude and the gap decrease with increasing disorder due to enhancement of Coulomb repulsion, and both fall to zero at the SIT. However, Finkel'stein's analysis assumes that the pairing amplitude is uniform across the system.

BdG calculations show that as V approaches V_c, the pairing amplitude $\Delta(\mathbf{r})$ becomes strongly inhomogeneous, as shown in Fig. 5. The superconductor breaks up into puddles, on the scale of the coherence length, where $\Delta(\mathbf{r})$ is large; in the surrounding regions, $\Delta(\mathbf{r})$ is negligible. Such puddles still exist even at large V.

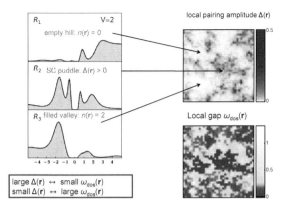

Fig. 7. (Right) Within BdG, the local pairing amplitude is *anticorrelated* with the local gap. (Left) LDOS results from QMC+MEM. Site R_1 is on a high potential hill that is nearly empty, and R_3 is in a deep valley that is almost doubly occupied. This leads to the characteristic asymmetries in the LDOS for R_1 and R_3. The small local pairing amplitude $\Delta(R)$ at these two sites is reflected in the absence of coherence peaks in the LDOS. In contrast, site R_2 has a density closer to half-filling, leading to a significant local pairing amplitude, a much more symmetrical LDOS, and coherence peaks that persist even at strong disorder.

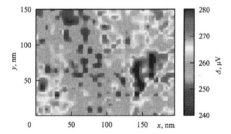

Fig. 8. Gap map of a TiN film obtained from scanning tunneling spectroscopy (from Sacepe et al., 2011) [27], showing inhomogeneities on a scale of a few tens of nanometers.

4.4. *Pairing amplitude versus single-particle gap*

Meanwhile, the single-particle gap E_{gap} remains finite (see Figs. 5 and 6). This is *not* an artifact of BdG; it is a very robust conclusion that is confirmed by quantum Monte Carlo (QMC) combined with maximum entropy methods (MEM) to extract spectral behavior. [c]

---

[c] One might think that Griffiths-McCoy-Wu rare regions might produce subgap weight. We have found that this effect is insignificant, and in any case, in two dimensions they do not affect the

Figure 7 shows that there is in fact an anti-correlation between the local pairing amplitude and the local spectral gap. The SC puddles on which the local $\Delta(\mathbf{R})$ is finite have a finite gap with symmetric line shapes and sharp coherence peaks or pile-ups in the density of states at the gap edges. On the other hand, the insulating regions have $\Delta(\mathbf{R}) \approx 0$ and very asymmetric broad density of states showing a much larger gap. Although the local gap extracted from the local density of states (LDOS) is highly inhomogeneous, it is nevertheless finite at every site, similar to the experimental data in Fig. 8.

The DOS is the LDOS averaged over all sites. The gap in the DOS, E_{gap}, is the lowest gap in the LDOS on any site. According to BdG (Fig. 5) and QMC (Figs. 6, 11) calculations, E_{gap} remains robust across the SIT, even when thermal and quantum phase fluctuations are included. Thus the SIT is a transition from a gapped superconductor to a gapped insulator.

The key reason for the robustness of the spectral gap even for high disorder is because of the disorder-induced "emergent granularity". The formation of blobs with finite local pairing amplitude leads to regions with a finite local gap. Further the spectra on these locally SC blobs are fairly symmetric because of strong number fluctuations and particle hole mixing. These regions are separated from insulating seas where the pairing amplitude is almost zero but not the local gap. The insulating regions arise form either deep valleys that are filled or high hills that are empty as seen in Fig. 7. Given the almost fixed number of particles (either two or zero) these regions allow for very small number fluctuations and hence considerably enhanced phase fluctuations of the conjugate variable leading to an almost zero pairing amplitude. The gap in these regions is the energy difference between the chemical potential and the local energy in the valley or the hill. This gap is considerably larger than that on the SC blobs and shows an asymmetric line shape because of the lack of particle-hole mixing.

We thus see that all approaches (atomic limit, pairing of exact eigenstates, BdG, and QMC) concur on the existence of a robust, finite gap in the single-particle density of states.

4.5. *Pseudogap over wide temperature range*

The hard gap at $T = 0$ evolves into a pseudogap – a suppression in the low-energy DOS – which persists well above the superconducting T_c up to a crossover temperature scale T^*, in marked deviation from BCS theory. This disorder-driven pseudogap also exists at finite temperatures in the insulating state and grows with disorder (Fig. 9). These predictions are in good agreement with experiments[27,28].

---

conclusion of a robust single-particle gap. Rare region effects will be dealt with in a forthcoming paper.

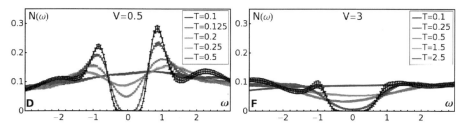

Fig. 9. Temperature dependence of DOS from QMC+MEM calculations. (Top) At weak disorder, as a function of increasing temperature, thermal fluctuations destroy the coherence peaks for $T \gtrsim T_c \approx 0.14$. However, a pseudogap remains up to higher temperatures $T \sim 0.4$. (Bottom) At strong disorder, there are no coherence peaks; there is a hard gap at $T = 0$ and a pseudogap up to $T \sim 1.5$.

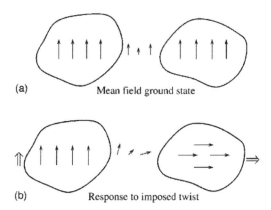

Fig. 10. The superfluid density is a measure of the rigidity of the phases. This rigidity is clearly reduced by thermal fluctuations. It is also reduced by amplitude variations and by quantum phase fluctuations even at $T = 0$. The upper panel shows a phase coherent ground state albeit with variations in the amplitude– large values in the SC puddles and small values in the intervening sea. For an applied twist, the system can accommodate most of the twist in regions where the amplitude is small leading to a very small cost in energy and hence a very small superfluid density.

5. Indicators of the phase-fluctuation-driven SIT: superfluid stiffness and off-diagonal long-range order

If the pairing amplitude and gap are both finite at all V, and the single-particle states are all localized, then what is the mechanism of the SIT?

Our BdG and QMC calculations[19] imply that the SIT is driven by *phase fluctuations*. For low disorder the local pairing amplitude $\Delta(\mathbf{R}) \equiv \langle c_{\mathbf{R}\downarrow} c_{\mathbf{R}\uparrow} \rangle$ is homogeneous across the system. However, as disorder is increased the system self-organizes into superconducting blobs on the scale of the coherence length within an insulating matrix (as seen in Fig. 5). The phases of the different blobs are coupled by Josephson tunneling of pairs. In the globally superconducting state, the phases of the different blobs get locked together whereas in the insulator the phase coherence of the different blobs is lost on ever shorter length and time scales as one moves away from the quantum phase transition. This is illustrated schematically in Fig. 10.

5.1. *Superfluid stiffness*

Ultimately, the SIT is defined by off-diagonal long-range order (ODLRO). ODLRO manifests itself in the two-particle correlator – i.e., the amplitude of inserting a pair and removing it at a different time and place. In other words, ODLRO means that pairs are delocalized and phase coherent, and its absence means that pairs are localized and incoherent.

We have not actually calculated the ODLRO correlator in BdG or QMC. Rather, we focused on an experimentally measurable quantity, the superfluid stiffness $\rho_s \propto D_s$. This is determined by the current-current correlator, and it is also a reliable indicator of superconducting long-range order.

The left panel of Fig. 11 shows the superfluid stiffness according to BdG (dashed curve) and as renormalized within the self-consistent harmonic approximation (SCHA, solid curve). D_s falls to zero at $V = V_c \sim 1.6t$ (for the given parameters). The right panel shows results from QMC. The superfluid stiffness is finite for $V < V_c$, and zero for $V > V_c$; in fact, the QMC simulations give superconductivity in a region of the phase diagram, $T < T_c(V)$.

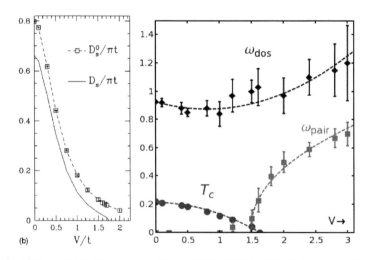

Fig. 11. (Left) Superfluid stiffness according to BdG+SCHA. (Right) Energy and temperature scales across the superconductor-insulator transition (SIT) according to QMC+MEM calculations. The single-particle gap ω_{dos} remains finite for all values of disorder V, whereas the superconducting T_c and the two-particle energy scale ω_{pair} in the insulator both vanish at the SIT.

5.2. Pair susceptibility

As seen form Fig. 11, the transition temperature T_c is suppressed to zero at the quantum phase transition $V = V_c$. What is the energy scale on the insulating side that vanishes at the transition? We discover the answer in the properties of the two-particle transport $P(\mathbf{r}, \mathbf{r}', \tau)$. The local two-particle spectral function, or pair susceptibility $P(\omega)$, is defined as the analytic continuation of the correlation function $P(\tau) = \sum_\mathbf{R} \langle T_\tau F(\mathbf{R}; \tau) F^\dagger(\mathbf{R}; 0) \rangle$ where $F(\mathbf{R}, \tau) = c_{\mathbf{R}\downarrow}(\tau) c_{\mathbf{R}\uparrow}(\tau)$. Physically, $P(\mathbf{R}, \omega)$ is the amplitude for inserting a pair at a site $\mathbf{R}$ at energy ω, and $P(\omega)$ is the average insertion amplitude over all sites.

Fig. 12 shows QMC+MEM results for the imaginary part of P. On the superconducting side of the transition there is a large amplitude for inserting pairs at zero energy. However, on the insulating side, there is a characteristic energy scale ω_{pair} to insert a pair in the insulator that collapses upon approaching the SIT (notwithstanding a small amount of spectral weight at low energies coming from rare regions).

Thus the the local two-particle energy scale, ω_{pair}, is an indicator of the approach to the SIT on the *insulating* side. A finite value of ω_{pair} tells us about the time scales ($\approx \omega_{\text{pair}}^{-1}$) over which the pairs are able to travel coherently through the system. As the transition is approached from the insulating side these time scales get longer and ultimately diverge at the transition.

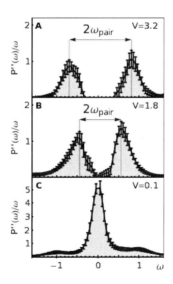

Fig. 12. Imaginary part of the dynamical pair susceptibility $P''(\omega)/\omega$ at $T = 0.1t$. Error bars represent variations between 10 disorder realizations. For $V < V_c$ the large peak at $\omega = 0$ indicates zero energy cost to insert a pair into the SC. For $V > V_c$, there is a gap-like structure at $\pm\omega_{\text{pair}}$, the typical energy required to insert a pair into the insulator.

6. Conclusions

In two dimensions, the metal-insulator transition for the underlying non-interacting disordered problem occurs at $V = 0^+$, that is all single-particle states are localized at any finite V. In contrast, the SIT for infinitesimal $|U|$ occurs at finite V. Thus even with all single-particle states localized, "long-distance two-particle transport is possible" because of the presence of the condensate in which pairs are delocalized and coherent". Our main results are summarized in Fig. 11. The superconducting phase persists up to a finite temperature $T_c(V)$, shown in Fig. 11. Beyond that a pseudogap persists in the density of states though coherence peaks are suppressed at T_c. In this regions there are signatures of pairing but thermal and quantum fluctuations destroy phase coherence. The insulator is a novel kind of insulator possessing a finite gap to single-particle excitations, which indicates that it consists of bound pairs. The size of these pairs should be viewed to be on the size of the local SC blobs. However, in spite of the presence of these pairs or blobs the system is not a SC at a global level because the phases on the different blobs are phase incoherent due primarily to quantum phase fluctuations in the proximity of the quantum phase transition.

With these calculations we have nailed the nature of the phases and the mechanism of the disorder-driven quantum phase transition. The next level of open questions are now related to the behavior of the frequency-dependent response functions and transport as a function of temperature and disorder across the SIT.

Acknowledgments

We gratefully acknowledge support from US Department of Energy, Office of Basic Energy Sciences grant DOE DE-FG02-07ER46423 (N.T., Y.L.L.), NSF DMR-0907275 (K.B.), NSF DMR-1006532 (M.R.), and computational support from the Ohio Supercomputing Center.

References

1. P. W. Anderson, *Journal of Physics and Chemistry of Solids* **11**, 26 (September 1959).
2. M. Ma and P. A. Lee, *Phys. Rev. B* **32**, 5658 (Nov 1985).
3. M. Strongin, R. S. Thompson, O. F. Kammerer and J. E. Crow, *Phys. Rev. B* **1**, 1078 (Feb 1970).
4. D. B. Haviland, Y. Liu and A. M. Goldman, *Phys. Rev. Lett.* **62**, 2180 (May 1989).
5. J. M. Valles, R. C. Dynes and J. P. Garno, *Phys. Rev. B* **40**, 6680 (October 1989).
6. J. M. Valles, R. C. Dynes and J. P. Garno, *Phys. Rev. Lett.* **69**, 3567 (December 1992).
7. A. F. Hebard and M. A. Paalanen, *Phys. Rev. Lett.* **65**, 927 (Aug 1990).
8. D. Shahar and Z. Ovadyahu, *Phys. Rev. B* **46**, 10917 (Nov 1992).
9. J. A. Chervenak and J. M. Valles, *Phys. Rev. B* **59**, 11209 (May 1999).
10. M. A. Steiner, G. Boebinger and A. Kapitulnik, *Phys. Rev. Lett.* **94**, 107008 (Mar 2005).
11. M. D. Stewart, A. Yin, J. M. Xu and J. M. Valles, *Science* **318**, 1273 (Nov 2007).
12. H. Q. Nguyen, S. M. Hollen, M. D. Stewart, J. Shainline, A. Yin, J. M. Xu and J. M. Valles, *Phys. Rev. Lett.* **103**, 157001 (October 2009).

13. Y. Lee, C. Clement, J. Hellerstedt, J. Kinney, L. Kinnischtzke, X. Leng, S. D. Snyder and A. M. Goldman, *Phys. Rev. Lett.* **106**, 136809 (Apr 2011).
14. Y.-H. Lin and A. M. Goldman, *Phys. Rev. Lett.* **106**, 127003 (Mar 2011).
15. A. T. Bollinger, G. Dubuis, J. Yoon, D. Pavuna, J. Misewich and I. Bozovic, *Nature* **472**, 458 (April 2011).
16. S. Sachdev, *Quantum Phase Transitions* (Cambridge, London, 1999).
17. A. Ghosal, M. Randeria and N. Trivedi, *Phys. Rev. Lett.* **81**, 3940 (November 1998).
18. A. Ghosal, M. Randeria and N. Trivedi, *Phys. Rev. B* **63**, 020505 (December 2000).
19. K. Bouadim, Y. L. Loh, M. Randeria and N. Trivedi, *Nature Physics* **7**, 884–889 (2011).
20. E. Abrahams, P. W. Anderson, D. C. Licciardello and T. V. Ramakrishnan, *Phys. Rev. Lett.* **42**, 673 (Mar 1979).
21. P. Lee and T. V. Ramakrishnan, *Rev. Mod. Phys.* **57**, 287 (1985).
22. A. MacKinnon and B. Kramer, *Phys. Rev. Lett.* **47**, 1546 (November 1981).
23. M. V. Feigel'man, L. B. Ioffe, V. E. Kravtsov and E. A. Yuzbashyan, *Phys. Rev. Lett.* **98**, 027001 (Jan 2007).
24. M. Feigel'man, L. Ioffe, V. Kravtsov and E. Cuevas, *Annals of Physics* **365**, 1368 (2010).
25. I. Burmistrov, I. Gornyi and A. Mirlin, *Phys. Rev. Lett.* **108**, 017002 (2012).
26. A. M. Finkel'stein, *Physica B* **197**, 636 (1994).
27. B. Sacépé, T. Dubouchet, C. Chapelier, M. Sanquer, M. Ovadia, D. Shahar, M. Feigel'man and L. Ioffe, *Nat. Phys.* **7**, 239 (March 2011).
28. M. Mondal, A. Kamlapure, M. Chand, G. Saraswat, S. Kumar, J. Jesudasan, L. Benfatto, V. Tripathi and P. Raychaudhuri, *Phys. Rev. Lett.* **106**, 047001 (Jan 2011).

THE SPIN GLASS-KONDO COMPETITION IN DISORDERED CERIUM SYSTEMS

S. G MAGALHAES[1*], F. ZIMMER[2] and B. COQBLIN[3,†]

[1] *Instituto de Fisica, Universidade Federal Fluminense, Niteroi, 24210 346, RJ, Brazil*
[2] *Dep. de Fisica, Universidade Federal de Santa Maria, Santa Maria, 97105 900, RS, Brazil*
[3] *Laboratoire de Physique des Solides, CNRS, Universite Paris-Sud, 91405-Orsay, France*
*sgmagal@gmail.com

We discuss the competition between the Kondo effect, the spin glass state and a magnetic order observed in disordered Cerium systems. We present firstly the experimental situation of disordered alloys such as $CeNi_{1-x}Cu_x$ and then the different theoretical approaches based on the Kondo lattice model, with different descriptions of the intersite exchange interaction for the spin glass. After the gaussian approach of the Sherrington-Kirkpatrick model, we discuss the Mattis and the van Hemmen models. Then, we present simple cluster calculations in order to describe the percolative evolution of the clusters from the cluster spin glass to the inhomogeneous ferromagnetic order recently observed in $CeNi_{1-x}Cu_x$ disordered alloys and finally we discuss the effect of random and transverse magnetic field.

Keywords: Spin glasses, Kondo Lattice models, disordered Cerium systems.

1. Introduction

The influence of the disorder on the behavior of strongly correlated electron systems has been extensively studied and we will discuss here the strong competition between the Kondo effect, a magnetic order and the spin glass state occurring in disordered Cerium systems. In fact, there are only a few systems which present together the Kondo effect and a spin glass behavior with also a magnetic order. Such a coexistence has been observed in disordered Cerium or Uranium alloys and we will firstly discuss the experimental situation of these systems.

First, the $CeNi_{1-x}Cu_x$ alloys have been extensively studied and their properties vary with the concentration of Nickel. For large x values above 0.7, the alloys are antiferromagnetic at low temperatures, while, for small x values below 0.2, they present a Kondo behavior. But, for $0.3 < x < 0.7$, these alloys change with decreasing temperature from the paramagnetic state to a spin glass one and finally to a ferromagnetic order.[1,2] There has been recently a considerable progress in the experimental knowledge of the phase diagram and magnetic clusters occur around

[†] In memoriam.

the spin glass-ferromagnetic transition; this new phase diagram will be described in the section 3 introducing some new calculations on magnetic clusters.[3-5]

Another example is provided by the $CeRh_xPd_{1-x}$ alloys which are clearly ferromagnetic for x values smaller than 0.65 and present a complicate Kondo behavior for large x values. The Curie temperature disappears for x around 0.8, but there is no quantum critical point. Instead, a peculiar "Kondo-cluster-glass" is found and Non-Fermi-Liquid effects in the specific heat, ac susceptibility and magnetization support a spin glass state described by the Griffiths model.[6,7]

The alloys $Ce_2Au_{1-x}Co_xSi_3$ present also a peculiar phase diagram: for small x values below 0.4, an antiferromagnetic phase is followed with decreasing temperature by a spin glass phase; then there exists only an antiferromagnetic phase up to roughly x=0.8 and finally a Kondo lattice state occurs for very large x values tending to 1.[8]

The compound $CeAgGa$ shows a spin-glass state with a freezing temperature of 5.1 K and a ferromagnetic-like order at the temperature of 3.6 K.[9] The spin glass state appears to arise from a random distribution of the Ag and Ga atoms around the Ce atoms and this compound looks like the $CeNi_xCu_{1-x}$ alloys for an intermediate concentration. The compound Ce_2NiGe_3 is also a Kondo lattice compound showing a spin glass behavior.[10] In fact, there are often some complicate behaviors involving a Non-Fermi-Liquid effect rather than the classical Kondo effect. For example, in the $CeCoGe_{3-x}Si_x$ alloys (with $0 < x < 3$), the phase diagram shows, with increasing x, an antiferromagnetic order, then a Kondo Fermi Liquid behavior and finally a Non Fermi Liquid (NFL) behavior.[11] But, above 6.5 kbar, the compound $CeCoGe_{2.2}Si_{0.8}$ undergoes a change to a spin-glass like behavior.[12]

A similar situation occurs with alloys containing Uranium instead of Cerium. Different behaviors such as antiferromagnetic order, NFL, spin glass and superconductivity have been observed in $U_{1-x}R_xPd_2Al_3$ alloys where R can be La, Th or Y;[13] in the case of $U_{1-x}La_xPd_2Al_3$ alloys, one observes first an antiferromagnetic order with a decreasing Neel temperature for x increasing from 0 to 0.2, then a spin glass phase up to $x = 0.7$ and finally a Kondo state with a NFL behavior. Similar properties have been shown in $UCu_{5-x}Pd_x$.[14,15] The phase diagram of $UCu_{5-x}Pd_x$ alloys presents an antiferromagnetic order for an increasing concentration x of order 0.8 and a spin glass behavior for x larger than 1.5. For x between 1 and 1.5, there is occurrence of a NFL behavior certainly connected to the Kondo effect.[15] The NFL properties of these Uranium systems have been also extensively studied.[16,17] In fact, the competition between the Kondo effect and the spin glass in Uranium systems needs a treatment of the Kondo effect within the Underscreened Kondo Lattice model[18] and we will not discuss here the case of Uranium disordered alloys.

We have done a brief experimental review, but indeed details and especially the different phase diagrams can be found in the papers cited in reference. Now, after this brief review, we will present in the next section the basic models which have accounted for the experimental results, then in the following section the derivation of some simple calculations which can describe the clusters occurring in both the spin glass and the ferromagnetic phases and finally in the last section the effect of random and transverse magnetic field.

2. Models for the Kondo-Spin Glass-Magnetic Order Competition

There are different works on the competition between the two well known Kondo and spin glass effects,[19–21] but here we will present the models that we have recently developed to account for the phase diagrams of disordered Cerium alloys showing the competition between a spin glass phase, a Kondo state and a magnetic order, as observed in $CeNi_{1-x}Cu_x$ alloys. The previous disordered alloy is a very good experimental example to be accounted by our models: the compound CeNi has a Kondo behavior and the compound CeCu has a simple magnetic behavior; when the concentration x decreases from 1 to 0, the importance of the Kondo effect increases and consequently the intrasite Kondo coupling J_K increases. Thus, we can compare the different theoretical models with increasing J_K (as shown in Figure 1) to the experimental phase diagram observed in $CeNi_{1-x}Cu_x$ alloys with decreasing the concentration x.

To describe this behavior, we have taken first a conduction band and localized spins on each Cerium site; then we consider a Kondo interaction between the localized spin $\mathbf{S}_i$ and the spin $\mathbf{s}_i$ of a conduction electron at the same site. Then we add an intersite interaction between nearest-neighboring localized spins in order to describe the spin glass effect. In fact, it is well known that the intersite interaction depends on the model chosen to describe the spin glass physics and we will review very briefly the first models used and we will present different comparisons with experiment. Thus, we have performed different calculations for the spin glass-Kondo interaction and the main difference between them lies in the approach used to describe the spin glass.

We start from the following general Hamiltonian :

$$H = \sum_{\mathbf{k}\sigma} \varepsilon_{\mathbf{k}} n^c_{\mathbf{k}\sigma} + \sum_{i\sigma} E_0 n^f_{i\sigma} + J_K \sum_i \mathbf{s}_i \cdot \mathbf{S}_i + \sum_{i,j} J_{ij} S^z_{fi} S^z_{fj} \qquad (1)$$

where $\varepsilon_{\mathbf{k}}$ is the energy of the conduction band, $J_K > 0$ is the intrasite Kondo coupling and J_{ij} is the intersite interaction used to describe the spin glass effect.

In fact, the disorder is described by the interaction between the spins $\mathbf{S}_i$ of different magnetic Cerium atoms, as it is the case in spin-glass systems and the randomness cannot come from the intrasite Kondo effect, but only from intersite interactions. Also, we have taken here the Ising model which was already used in many spin glass calculations. In fact, the ground state is often a doublet in Cerium systems, which gives a good justification for the use of the simpler Ising model.

To describe the nature of the spin glass, we start firstly with the Sherrington-Kirkpatrick model[22] in which the couplings J_{ij} are taken here as independent random variables described by a gaussian distribution :

$$\mathcal{P}(J_{ij}) = \frac{1}{J}\sqrt{\frac{N}{64\pi}} \exp\left\{-\frac{(J_{ij} + 2J_0/N)^2}{64J^2} N\right\}. \qquad (2)$$

In the preceding expression, the average value J_0 has been taken firstly equal to zero[23] and this calculation describes the SG-Kondo transition. In order to try to obtain a comparison with experiment, we have considered the case $J_0 < 0$, which produces a complex phase diagram with spin glass, ferromagnetic, mixed (a spin glass with spontaneous magnetization) and Kondo phases[24] and the case $J_0 > 0$, which corresponds to an antiferromagnetic phase in competition with Spin Glass and Kondo phases.[25] Another calculation has been done in order to include the effect of a transverse magnetic field.[26]

In these calculations, we have taken the Sherrington-Kirkpatrick model with a gaussian distribution of the intersite exchange integrals.[22] The first studied case takes $J_0=0$[23] and we discuss the results as a function of only the 2 parameters J and J_K characteristic of the spin glass and of the Kondo states. The functional integration techniques and the mean field approximation are used here ; to describe the Kondo effect, we use the correlation $\lambda_\sigma = <c^+_{i\sigma} f_{i\sigma}>$ between conduction and f electrons. The resulting phase diagram in the $T - J_K$ plane for a fixed J value gives a spin glass phase for small J_K values and a Kondo phase for large J_K values. But in fact, the experiments on $CeNi_{1-x}Cu_x$ alloys show clearly a additional competition with a ferromagnetic order. Then, in order to obtain a more complex phase diagram with a ferromagnetic[24] or an antiferromagnetic[25] phase occurring at low temperatures for small J_K values, we have taken the same model with a non zero value J_0 of the center of the gaussian distribution. Thus, it results that the Kondo phase obtained for large J_K values is still there, but that there is a competition between the spin glass phase and the magnetic phase for smaller J_K values. However, in the ferromagnetic case, when J_K is typically of order J_0, we have obtained, with decreasing temperature, successively a ferromagnetic phase, then a mixed ferromagnetic-spin glass phase and finally a spin glass phase. The occurrence of a spin glass, Kondo and ferromagnetic phases in the same phase diagram is a good result, but the evidence of a spin glass phase at very low temperatures below the ferromagnetic one is in disagreement with experimental results previously obtained for $CeNi_{1-x}Cu_x$ alloys.

In order to improve the agreement with experiment, we have more recently developed a generalization of the Mattis model[27] and of the van Hemmen model[28] which both describe the bonds J_{ij} joining the localized spins as depending on a random choice on each atom. We have developed firstly a generalization of the Mattis model,[27] which represents an interpolation between ferromagnetism and a highly disordered spin glass.[29]

In the Mattis model,[27] which has been proposed as a solvable model to the spin glass problem, the bonds joining the localized spins have been defined as separable random variables ξ_i. In this model the couplings between spins are given by :

$$J_{ij} = \frac{1}{N} \sum_{\mu\nu} J_{\mu\nu} \xi_i^\mu \xi_j^\nu, \tag{3}$$

where $\xi_i^\mu = \pm 1$ ($\mu = 1, 2, ..., p$; $i = 1, 2, ..., N$) are independent random distributed variables. For the classical Ising model, if $\mu = \nu = 1$, the original Mattis model[27] is recovered. However, if $J_{\mu\nu} = J\delta_{\mu\nu}$ and $p - N$ with the N^2 random variables ξ_i^μ having mean zero and variance one, in the limit of N large J_{ij} tends to a Gaussian variable with mean zero and variance $N^{-1/2}J$ as in the SK model.[22] Therefore, we can consider this model as an interpolation between ferromagnetism and highly disordered spin glass.[30]

For a reasonable value of the relative importance of the ferromagnetic and spin glass phases, i.e. for a relatively small J_K/J ratio, we have obtained, with decreasing temperature, a spin glass phase, then a mixed ferromagnetic-spin glass phase and finally a ferromagnetic phase. The Kondo phase is always present for large J_K/J. This theoretical result accounts better for the experimental phase diagram of $CeNi_{1-x}Cu_x$ alloys., with in particular at low temperatures the ferromagnetic phase below the spin glass phase.

Finally, we have used recently the van Hemmen model[28] to describe the interplay of Kondo effect, Spin Glass and Ferromagnetism in disordered Ce alloys. This model gives a good description of this interplay and appears to be simpler than the previous models, in particular for the treatment of clusters, as we will see later on.[31]

Thus, in order to improve again the agreement with the experimental case of $CeNi_{1-x}Cu_x$ alloys, we introduce here a new kind of disordered coupling J_{ij} given by van Hemmen (vH)[28] as:

$$J_{ij} = \frac{J}{N}(\xi_i\eta_j + \eta_i\xi_j) + \frac{I_0}{N} \quad (4)$$

where ξ_i and η_j are equal to ± 1 and are random variables which follow a bimodal distribution.[31] We take here both a random SG contribution and a ferromagnetic one respectively proportional to the parameters J and I_0. Our calculations using the vH approach have given different phase diagrams which depend on the ratio I_0/J. Several cases are described in Ref.[31] and we have obtained the typical case of a transition with decreasing temperature from spin glass (SG) to ferromagnetism (FE) with an intermediate "mixed" phase SG+FE for small J_K values and still the Kondo phase for large J_K values.[31] The use of the van Hemmen model offers some advantages: first the calculation does not need the use of the replica method and is simpler; second there exists a real mixed phase between Spin Glass and Ferromagnetism, which is in good agreement with the experimental phase diagram of $CeNi_{1-x}Cu_x$ alloys,[31] as shown on Figure 1.

Our recent work, with the Kondo effect, ferromagnetism and the van Hemmen model used for the spin glass, gives therefore a good description of the phase diagram of $CeNi_{1-x}Cu_x$ alloys,[31] but recent experimental progress has shown the occurrence of clusters and we will describe in the next section recent experimental and theoretical progress. However, in conclusion of this section, we can say that the more "local" description[29,31] given by the Mattis or the van Hemmen model seems

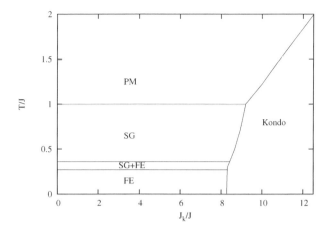

Fig. 1. Phase diagram $T/J - J_K/J$ for $J_0/J=1.4$.[31]

to be better here than the "average" description using the Sherrington-Kirkpatrick approach[23,24] to describe the phase diagram of $CeNi_{1-x}Cu_x$ alloys.

3. The Kondo Cluster — Glass Model

Up to now, we have discussed different approaches to describe the Kondo-spin glass-ferromagnetism competition and we have applied them successfully to the first description of the phase diagram of Cerium disordered alloys like $CeNi_{1-x}Cu_x$. But, more sophisticated experiments performed at very low temperatures have shown the presence of magnetic clusters in these alloys. More precisely, measurements of the ac susceptibility, dc magnetization and specific heat down to 0.1 K show that a cluster-glass state is formed at a freezing temperature T_f well above the ferromagnetic state. Recent Muon spin relaxation and Small Angle Neutron scattering measurements confirm the existence of magnetic clusters and of a percolative transition with decreasing temperature.[3] Thus, it is now established that, in $CeNi_{1-x}Cu_x$ alloys for x between 0.3 and 0.6, an inhomogeneous "cluster spin glass" (CSG) occurs below the paramagnetic regime and is followed, with decreasing temperature, by an inhomogeneous ferromagnetic (IFM) order, with a percolative evolution of the clusters from the CSG state to the IFM state.[4,5] Figure 2 summarizes the evolution of the cluster glass which does not show a phase transition, but a percolative evolution. A similar cluster spin glass has been observed in $CePd_{1-x}Rh_x$.[6,7]

We have developed several works to account for the cluster glass and the starting point is to consider small clusters with a constant size and with an intra-site Kondo interaction, an intra-cluster ferromagnetic interaction and an inter-cluster random magnetic interaction. The first work uses the classical Sherrington-Kirkpatrick model to treat the spin glass interaction,[32] but more recently we have developed the more convincing van Hemmen model.[33]

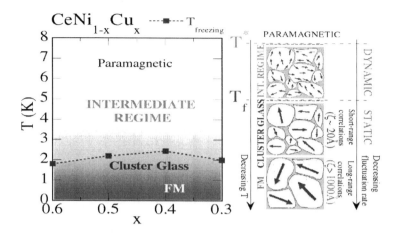

Fig. 2. Experimental phase diagram for $CeNi_{1-x}Cu_x$ alloys.[5]

To study the competition between the Kondo effect and the cluster spin glass, we use the following Hamiltonian:

$$H = \sum_{a=1}^{N_{cl}} \sum_{i=1}^{n_s} \epsilon_0 \sum_{\sigma=\uparrow\downarrow} \hat{n}_{i\sigma a}^f - J_0 \sum_{a=1}^{N_{cl}} \sum_{i<j}^{n_s} \hat{S}_{ia}^z \hat{S}_{ja}^z + \sum_{a}^{N_{cl}} \sum_{i,j} t_{ij}^{aa} d_{i\sigma a}^\dagger d_{j\sigma a} \quad (5)$$

$$+ J_K \sum_{a=1}^{N_{cl}} \sum_{i=1}^{n_s} \left(\hat{S}_{ia}^+ s_{ia}^- + \hat{S}_{ia}^- s_{ia}^+ \right) - \sum_{a<b}^{N_{cl}} J_{ab} \hat{S}_a^z \hat{S}_b^z$$

We have here $N = N_{cl}.n_s$, where N_{cl} and n_s are the number of clusters and the number of spins in each cluster, respectively. The hopping t_{ij}^{aa} is only inside the cluster. The indices (a, b) refer to clusters while (i, j) indicate spins inside a cluster. So, we can write $S_{ia}^+ = f_{ia\uparrow}^\dagger f_{ia\downarrow}$, $s_{ia}^- = d_{ia\downarrow}^\dagger d_{ia\uparrow}$ and

$$\hat{S}_a^z = \sum_{i=1}^{n_s} \hat{S}_{ia}^z = \sum_{i=1}^{n_s} \sum_{\sigma=\uparrow\downarrow} \sigma \hat{f}_{i\sigma a}^\dagger \hat{f}_{i\sigma a} \quad (6)$$

The intercluster coupling J_{ab} is a random variable given in the van Hemmen model by an equation similar to equation (4). Details on our first treatment using the Sherrington-Kirkpatrick model can be found in Ref. 32 and we discuss here only the van Hemmen (vH) model.[33]

This model starts from Kondo and ferromagnetic intra-cluster interactions and an inter-cluster vH interaction given by equation (4). The inter-cluster interactions are treated in a mean field level without the use of replica method, in which the magnetization and spin glass order parameters are introduced in the partition function. These order parameters are chosen to minimize an effective one-cluster free energy and the resulting intra-cluster interactions are then obtained by exact diagonalization. For this purpose, it is assumed that the localized fermions can act on

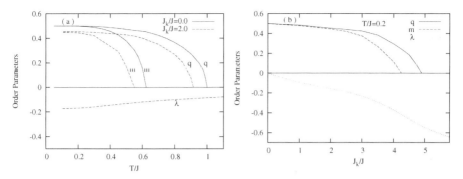

Fig. 3. (a) Order parameters q, m and λ as a function of T/J for $J_K/J = 0$ and 2. (b) q, m and λ versus J_K/J for $T/J = 0.2$. These results are for $n_s = 5$ and $I_0/J = 0.1$.[33]

a basis of states where each site is exactly occupied by only one fermion, while the conduction fermions can act on states with four possible occupancies at each site (site unoccupied, single occupied and double occupied).

The number of atoms n_s inside each cluster is presently limited to 4 or 5.[33] The parameters λ, q and m design respectively the Kondo order parameter, the spin glass order parameter and the magnetization. However, for relatively large values of J_K and I_0, we have obtained successively, with decreasing temperature, a Kondo phase followed by a Kondo cluster spin glass where both the Kondo and the cluster spin glass (CSG) order parameters are non zero; finally a mixed CSG and ferromagnetic phase is obtained. The results are shown in Figure 3 corresponding to clusters with a number $n_s = 5$ of atoms. The Kondo correlation is increasing with J_K as usual, and the figure shows a competition between the Kondo effect and the magnetic orders, the cluster spin glass (CSG) and the mixed phase (CSG+FE). These magnetic orders are strongly dependent on the ferromagnetic inter-cluster interaction I_0 and J_K.[33] It is important to remark that the results depend on the size of the cluster if we compare the results obtained for $n_s = 5$ to those obtained for $n_s = 4$.[33] We have performed here calculations with a fixed number of atoms per cluster and unfortunately we cannot at present do a calculation with an increasing cluster size with decreasing temperature, as observed experimentally. However, such sequence of phases can present a first account for the experimental percolative CSG-IFM transition observed in disordered Cerium alloys.[5,33]

4. The Effect of a Random and Transverse Magnetic Field

The last calculation that we have recently performed concerns the different effects of random fields on the global phase diagram already obtained in Ref. 23. The main motivation of this new type of disorder is that for intermediate doping in the $CeNi_{1-x}Cu_x$ alloys, the hysteresis curve exhibits cycles with sharp macroscopic jumps in the magnetization at low temperatures.[35] Recently, these jumps have been attributed to the presence of random anisotropies.[35] When these anisotropies are very strong, they can be related with the presence of random fields.[34] Therefore, it is added to the Hamiltonian given in Eq.1 a random field h_i following a gaussian distribution as shown below:

$$P(h_i) = \left[\frac{1}{2\pi\Delta^2}\right]^{1/2} \exp\left[-\frac{h_i^2}{2\Delta^2}\right]. \qquad (7)$$

The results found for the phase diagram[37] indicate that the random field can disrupt all magnetic long range orderings including the SG phase. Moreover, at lower temperature, it is also found that it is necessary to take larger values of the Kondo interaction in order to find a Kondo solution. In that sense, we can say that the presence of a random field also suppresses the Kondo singlet formation. Another interesting result related to the presence of a random field is the coexistence of the Kondo state and the so called independent spin phase. This phase represents a magnetic state in which the localized spins, even remaining uncorrelated, are oriented randomly since they couple locally with the random field.[38]

5. Conclusion

We have discussed in this work the competition between Kondo effect, magnetic order and spin glass which occurs in some Cerium disordered alloys. The experimental results have shown clearly that this competition can produce several nontrivial results as deviations of the Fermi liquid behaviour or cluster glass formation. From the theoretical point of view, the effects of disorder have been studied using the Kondo lattice model with an Ising intersite term by choosing several types of random distribution for the intersite spin coupling J_{ij}. This model has been also extended to include clusters of spins and finally, also random fields to provide a better description of the experimental scenario found in, for instance, $CeNi_{1-x}Cu_x$ alloys.

Although there was a considerable amount of theoretical results found already for the competition between Kondo effect, magnetic order and spin glass, some issues still remain unsolved. For example, a formulation in terms of spins clusters instead of canonical ones able to explain the effects of random fields in the mentioned competition is not done. Another very interesting open question is a proper treatment of the Kondo effect within the Underscreened Kondo Lattice to study the interplay between Kondo effect and spin glass as observed in some Uranium disordered alloys.

Acknowledgments

S.G.M and F.M.Z. acknowledge the support of the Brazilian agency CNPq. S.G.M. also acknowledges the agency FAPERJ.

References

1. J.C. Gomez Sal, J. Garcia Soldevilla, J.A. Blanco, J.I. Espeso, J. Rodriguez Fernandez, F. Luis, F. Bartolome, Phys. Rev. B **56**, 11741 (1997).
2. J. Garcia Soldevilla, J.C. Gomez Sal, J.A. Blanco, J.I. Espeso, J. Rodriguez Fernandez, Phys. Rev. B **61**, 6821 2000).
3. N. Marcano, J.I. Espeso, J.C. Gomez Sal, J. Rodriguez Fernandez, J. Herrero-Albillos, F. Bartolome, Phys. Rev. B **71**, 134401 (2005).
4. N. Marcano, J.C. Gomez Sal, J.I. Espeso, J.M. De Teresa, P.A. Algarabel, C. Paulsen and J.R. Iglesias, Phys. Rev. Lett. **98**, 166406 (2007).
5. N. Marcano, S.G. Magalhaes, B. Coqblin, J.C. Gomez Sal, J.I. Espeso, F.M. Zimmer and J.R. Iglesias, J. of Physics: Conf. Series, **200**, 012111 (2010).
6. T. Westerkamp, M. Deppe, R. Kuchler, M. Brando, C. Geibel, P. Gegenwart, A.P. Pikul and F. Steglich, Phys. Rev. Lett. **102**, 206404 (2009).
7. M. Brando, T. Westerkamp, M. Deppe, P. Gegenwart, C. Geibel and F. Steglich, J. of Physics: Conf. Series, **200**, 012016 (2010).
8. S. Majumdar, F.V. Sampathkumaran, St. Berger,, M. Della Mea, H. Michor, E. Bauer, M. Brando, J. Hemberger and A. Loidl, Solid State Comm. **121**, 665 (2002).
9. J. Goraus, A. Slebarski, M. Fijalkowski and L. Haweleck, Eur. Phys. J. B **80**, 65 (2011).
10. D. Huo, J. Sakurai, T. Kuwai, Y. Isikawa and Q. Lu, Phys. Rev. B **64**, 224405 (2001).
11. D. Eom, N. Takeda and M. Ishikawa, J. Phys. Soc. Japan **75**, 093706 (2006).
12. J. Larrea, S.B. Paschen, M. Muller, J. Teyssier and H. Ronnow, presented at SCES2011 Conference, Cambridge, Aug.2011.
13. V.S. Zapf, R.P. Dickey, E.J. Freeman, C. Sirvent and M.B. Maple, Phys. Rev. B **65**, 024437 (2001).
14. B. Andraka and G.R. Stewart, Phys. Rev. B **47**, 3208 (1993).
15. R. Vollmer, T. Pietrus, H.v. Lohneyssen, R. Chau and M.B. Maple, Phys. Rev. B **61**, 1218 (2000).
16. M.B. Maple, C.L. Seaman, D.A. Gajewski, Y. Dalichaouch, V.B. Barbetta, M.C. de Andrade, H.A. Mook, H.G. Lukefahr, O.O. Bernal and D.E. MacLaughlin, J. of Low Temperature Physics, **95**, 225 (1994).
17. M.B. Maple, M.C. de Andrade, J. Herrmann, Y. Dalichaouch, D.A. Gajewski, C.L. Seaman, R. Chau, R. Movshovich, M.C. Aronson and R. Osborn, J. of Low Temperature Physics, **99**, 223 (1995).
18. N.B. Perkins, M.D. Nunez-Regueiro, J.R. Iglesias and B. Coqblin, Phys. Rev. B **76**, 125101 (2007).
19. V. Dobrosavljevic, T.R. Kirkpatrick and G. Kotliar, Phys. Rev. Lett. **69**, 1113 (1992).
20. A.H. Castro Neto and B.A. Jones, Phys. Rev. B **62**, 14975 (2000).
21. E. Miranda and V. Dobrosavljevic, Rep.Progr. Phys., **68**, 2337 (2005).
22. D. Sherrington and S. Kirkpatrick, Phys. Rev. Lett. **35**, 1792 (1975).
23. Alba Theumann, B. Coqblin, S. G. Magalhaes and A.A. Schmidt, Phys. Rev. B **63**, 054409 (2001).
24. S. G. Magalhaes, A. A. Schmidt, Alba Theumann and B. Coqblin, Eur. Phys. J. B **30**, 419 (2002).
25. S. G. Magalhaes, A. A. Schmidt, F. M. Zimmer, Alba Theumann and B. Coqblin, Eur. Phys. J. B **34**, 447 (2003).
26. Alba Theumann and B. Coqblin, Phys. Rev. B **69**, 214418 (2004).
27. D.J. Mattis, Phys. Lett. **56A**,421 (1977).
28. J.L. van Hemmen, Phys. Rev. Lett. **49**, 409 (1982).
29. S. G. Magalhaes, F.M. Zimmer and B. Coqblin, Phys. Rev. B **74**, 014427 (2006).

30. D.J. Amit, *Modelling Brain Function, The World of Attractor Neural Networks*, Cambridge University Press, Cambridge, England (1989)
31. S. G. Magalhaes, F.M. Zimmer and B. Coqblin, Phys. Rev. B **81**, 094424 (2010).
32. F.M. Zimmer, S. G. Magalhaes and B. Coqblin, Physica B **404**,2972 (2009)
33. F.M. Zimmer, S. G. Magalhaes and B. Coqblin, J. of Physics: Conference Series **273**, 012069 (2011)
34. K.H. Fisher and J.A. Hertz, *Spin glasses*, Cambridge University Press, Cambridge, England (1991)
35. J.R. Iglesias, J.I. Espeso, N. Marcano and J.C. Gomez Sal, Phys.Rev. B **79**, 195128 (2009)
36. S. G. Magalhães, C. A. Morais, F. D. Nobre, J. Stat. Mech **2011**, p07014 (2011).
37. S. G. Magalhaes, F. M. Zimmer, B. Coqblin, "Spin-glass freezing in Kondo-lattice compounds in the presence of a random and transverse magnetic fields" to be submited, (2012)
38. T. Schneider, E. Pytte, Phys. Rev. B **15**, 1519 (1977).

TRANSPORT VIA CLASSICAL PERCOLATION AT QUANTUM HALL PLATEAU TRANSITIONS

MARTINA FLÖSER

Institut Néel, CNRS and Université Joseph Fourier, B.P. 166, 25 Avenue des Martyrs, 38042 Grenoble Cedex 9, France

SERGE FLORENS

Institut Néel, CNRS and Université Joseph Fourier, B.P. 166, 25 Avenue des Martyrs, 38042 Grenoble Cedex 9, France

THIERRY CHAMPEL

Université Joseph Fourier Grenoble I / CNRS UMR 5493, Laboratoire de Physique et Modélisation des Milieux Condensés, B.P. 166, 38042 Grenoble, France

We consider transport properties of disordered two-dimensional electron gases under high perpendicular magnetic field, focusing in particular on the peak longitudinal conductivity $\sigma_{xx}^{\text{peak}}$ at the quantum Hall plateau transition. We use a local conductivity model, valid at temperatures high enough such that quantum tunneling is suppressed, taking into account the random drift motion of the electrons in the disordered potential landscape and inelastic processes provided by electron-phonon scattering. A diagrammatic solution of this problem is proposed, which leads to a rich interplay of conduction mechanisms, where classical percolation effects play a prominent role. The scaling function for $\sigma_{xx}^{\text{peak}}$ is derived in the high temperature limit, which can be used to extract universal critical exponents of classical percolation from experimental data.

PACS numbers: 73.43.Qt, 64.60.ah, 71.23.An

1. Introduction

The quantum Hall effect[1,2] in two-dimensional electron gases (2DEG) follows from a disorder-induced localization process peculiar to the situation of large perpendicular magnetic fields B. While the formation of discrete Landau levels (LL) at energies $E_n = \hbar\omega_c\left(n + \frac{1}{2}\right)$ can account for the existence of robust quantum numbers (with $\omega_c = |e|B/m^\star$ the cyclotron frequency, $e = -|e|$ the electron charge, $m^\star$ the effective mass, $\hbar$ Planck's constant divided by 2π, and n a positive integer), the existence of a macroscopic number of localized states in the bulk of 2DEG is an essential aspect of quantum Hall physics[3]. Many questions are yet still open thirty years after the initial discovery: i) for metrological purposes[4], which physical processes are limiting the plateau quantization of the Hall conductivity near the universal value $\sigma_{xy} = ne^2/h$?; ii) what is the nature of the localization/delocalization transition from one plateau to the next[5,6,7], whereupon highly dissipative transport sets in? Theoretically, the problem in its full complexity requires to understand the quantum dynamics of electrons subject to the Lorentz force and random local electric fields, possibly with the inclusion of dissipative processes such as electron-electron and electron-phonon interaction, that are sensitive issues when one considers transport properties.

So far, a lot of attention was turned towards the understanding of the delocalization process in terms of a zero-temperature quantum percolation phase transition, which still remains a challenge for the theory[5,6,7], despite intriguing experimental evidence from transport[8,9] and local scanning tunneling spectroscopy[10]. In this framework, the quantum tunneling and interference of the guiding center trajectories within a complex percolation cluster allow dissipation to develop in a non-trivial way at the quantum Hall transition. Obviously, increasing temperature from absolute zero will generate inelastic processes limiting the coherence between saddle points of the disorder landscape, so that the quantum character of the transition becomes progressively irrelevant. In that case, a simpler quasiclassical transport theory becomes valid[11,12,13,14], which incorporates the fast cyclotron motion with the slow guiding center drifting, and takes into account inelastic contributions to transport. The transport problem does not become however totally trivial, because classical percolation in the related advection-diffusion regime is still not fully understood[15].

The aim of the present paper is two-fold. First, we will show in Sec. 2 that high mobility samples display a very rich temperature behavior for the peak longitudinal conductivity $\sigma_{xx}^{\text{peak}}$ (at the plateau transition), leading to a complex succession of transport crossovers with universal powerlaws, see Fig. 1.

Second, we will present in Sec. 3 a general diagrammatic formalism[14] allowing to compute dissipative transport dominated by classical percolation effects in quantum

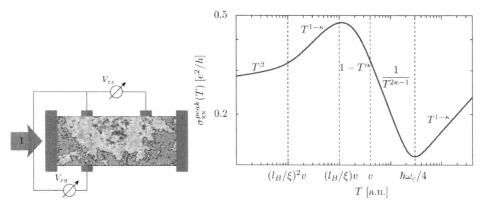

Fig. 1. Left: measurement of longitudinal V_{xx} and Hall V_{xy} voltages with applied current I in a two-dimensional sample with percolating random charge inhomogeneities. Right: sketch (on log-log scale) of the temperature dependence of the peak longitudinal conductivity $\sigma_{xx}^{\text{peak}}$ at the plateau transition. The existence of a hierarchy of energy scales (that are indicated by dashed lines) results in several crossovers between universal power-laws, as described later in the text.

Hall samples such as depicted in Fig. 1. In particular, we will be able to give strong support to a previously conjectured[11,12,13,15] critical exponent $\kappa = 10/13$ for the peak longitudinal conductivity $\sigma_{xx}^{\text{peak}}$ in the high temperature regime of the plateau transition. A universal scaling function describing the crossover for temperatures near the cyclotron energy (see the V-shaped part of the curve in Fig. 1) will be computed, giving a way to extract κ from experiments.

2. Classically percolating transport at the plateau transition

2.1. *Local conductivity model*

The starting point of transport calculations in the high temperature regime of the quantum Hall effect is a purely classical model[11,14], where the continuity equation $\nabla \cdot \mathbf{j} = 0$ (*i.e.* the continuum version of Kirchoff's law) is solved from the microscopic knowledge of Ohm's law $\mathbf{j}(\mathbf{r}) = \hat{\sigma}(\mathbf{r})\mathbf{E}(\mathbf{r})$, defining here the local conductivity tensor[16,17] that relates the local electric field to the local current density. Due to the existence of several energy scales, the disorder-induced spatial variations of $\hat{\sigma}(\mathbf{r})$ have a strong temperature dependence, which in turn affects the macroscopic transport properties. Drastic simplifications occur in the regime of high magnetic fields[17,18], where the combination of Lorentz force and local electrostatic potentials induces a slow drift motion in the direction orthogonal to the crossed magnetic and local electric fields. This vindicates the first simplification of the conductivity tensor

$$\hat{\sigma}(\mathbf{r}) = \begin{pmatrix} \sigma_0 & -\sigma_H(\mathbf{r}) \\ \sigma_H(\mathbf{r}) & \sigma_0 \end{pmatrix}, \quad (1)$$

where σ_0 encodes dissipative processes such as electron-phonon scattering, which will be assumed to be uniform in the bulk of the sample, and $\sigma_H(\mathbf{r})$ is the local Hall component, whose spatial dependence originates from charge density fluctuations due to disorder in the sample. We will discuss below the various regimes that can be expected for $\sigma_H(\mathbf{r})$ depending on the range of temperature T. For this purpose, we need to introduce the energy scales associated with the local disorder potential $V(\mathbf{r})$, and we define its typical amplitude $v = \sqrt{\langle [V(\mathbf{r})]^2 \rangle}$ and correlation length ξ. We will assume throughout that the disorder is smooth at the scale of the magnetic length $l_B = \sqrt{\hbar/eB}$ ($l_B = 8$nm at $B = 10$T), so that l_B/ξ is a small parameter. In what follows, a centered Gaussian distribution will be considered for the local electrostatic potential $V(\mathbf{r})$. The local conductivity model introduced here is valid at temperatures high enough so that phase-breaking processes, such as electron-phonon scattering, occur on length scales that are shorter than the typical variations of disorder. However, quantum mechanics may still be important to determine the microscopics of the conductivity tensor, as we argue below.

2.2. *Percolation effects in quantum Hall transport: phenomenology*

The occurence of percolation effects in the quantum Hall regime can be understood already from a quasiclassical perspective. In the high magnetic field limit, cyclotron and guiding center motions fully decouple, giving rise to Landau quantization on one hand, and to mainly closed trajectories of the guiding center on the other hand, that follow the equipotentials of the disorder landscape. Intuitively, the electrical current contributing to macroscopic transport will thus follow a percolation backbone. The crucial role of inelastic processes, controlled by the longitudinal component σ_0 in Eq. (1), can be understood by the fact that such current-carrying extended states must pass through many saddle-points on the disorder landscape. However, the drift velocity associated to the guiding center identically vanishes at these points, so that having a finite σ_0 is essential to connect the different valleys of the potential profile. The technical difficulty lies in evaluating the macroscopic conductivity in the limit where σ_0 is much smaller than the amplitude variations of the Hall component $\sigma_H(\mathbf{r})$, but yet does not fully vanish. This regime cannot simply be accessed from the $\sigma_0 \to 0$ limit, because the transport equation becomes singular. The strategy developped in Ref. 14 and Sec. 3 will be to extrapolate from high orders of the perturbatively controlled $\sigma_0 \to \infty$ expansion to the case of small dissipation.

Assuming that a critical state is established in the small σ_0 limit due to the scale invariant nature of the percolation backbone, one can infer from dimensional analysis that the macroscopic longitudinal conductivity scales as[11,14]:

$$\sigma_{xx} \propto \sigma_0^{1-\kappa}[\langle \sigma_H^2 \rangle - \langle \sigma_H \rangle^2]^{\kappa/2}, \qquad (2)$$

where κ is a non-trivial exponent previously conjectured[11,12,13,15] to be $\kappa = 10/13$, see Sec. 3 for a diagrammatic approach to this result. Based on simple microscopic arguments for the local Hall conductivity $\sigma_H(\mathbf{r})$ that we introduce now, it is possible to understand from Eq. (2) various transport regimes that are relevant for quantum Hall systems. In all what follows, we will assume that electron-phonon processes dominate in the longitudinal component[19], leading to the temperature dependence $\sigma_0(T) \propto T$.

2.3. A hierarchy of transport crossovers

2.3.1. Fully classical regime: $\hbar\omega_c \ll T$

At temperatures higher than the cyclotron energy, both cyclotron and drift motions are classical, so that the classical Hall's law prevails: $\sigma_H(\mathbf{r}) = (e/B)n(\mathbf{r})$, with $n(\mathbf{r})$ the local electronic density, which undergoes smooth spatial fluctuations in case of high mobility samples[11]. In relatively clean samples, the amplitude v of disorder fluctuations remain small compared to the classical cyclotron energy, so that the Hall conductivity follows at first order the spatial variations of the local potential:

$$\sigma_H(\mathbf{r}) = \frac{en}{B} + AV(\mathbf{r}). \qquad (3)$$

with n the total electron density and A a constant to be determined below. Thus, for a Gaussian distributed disorder, the local Hall conductivity displays Gaussian fluctuations and is weakly dependent on temperature. Using the percolation Ansatz (2) for the macroscopic longitudinal conductivity, we find:

$$\sigma_{xx} \propto v^\kappa T^{1-\kappa} \propto T^{3/13}, \qquad (4)$$

which shows already a first non-trivial behavior in temperature[12,14] connected to classical percolation, where the conductivity mildly decreases as temperature is lowered, see also Fig. 1.

2.3.2. *Formation of Landau levels:* $v \ll T \ll \hbar\omega_c$

As temperature crosses the cyclotron energy, Landau levels start to emerge, and the local density is given by Pauli's principle: $\sigma_H(\mathbf{r}) = \frac{e^2}{h} \sum_{m=0}^{\infty} n_F[E_m - V(\mathbf{r}) - \mu]$ with μ the chemical potential and $n_F(E) = 1/(e^{E/T} + 1)$ the Fermi-Dirac distribution (we set Boltzmann's constant $k_B = 1$ in what follows). We will neglect spin effects for simplicity in what follows (Landau levels are assumed spin non-degenerate). In the considered temperature range $v \ll T$, the Fermi distribution can be linearized, which leads to Eq. (3) with $A = (e^2/h)(\hbar\omega_c)^{-1}$ in the case $\hbar\omega_c \ll T$ considered previously, and more generally to:

$$\sigma_H(\mathbf{r}) = \frac{en}{B} + \frac{e^2}{h} \sum_{m=0}^{\infty} n'_F(E_m - \mu) V(\mathbf{r}). \qquad (5)$$

The local conductivity remains Gaussian, but acquires now an extra temperature dependence from the Fermi function, which can be illustrated in the case of the plateau $\nu \to \nu + 1$ transition, which leads for $T \ll \hbar\omega_c$ to

$$\sigma_H^{\text{peak}}(\mathbf{r}) = \frac{en}{B} + \frac{e^2}{h} \frac{1}{4T} V(\mathbf{r}). \qquad (6)$$

Using the percolation Ansatz (2), we find in the considered temperature range:

$$\sigma_{xx}^{\text{peak}} \propto \left(\frac{v}{T}\right)^\kappa T^{1-\kappa} \propto \frac{1}{T^{2\kappa-1}} \propto \frac{1}{T^{7/13}}, \qquad (7)$$

so that the peak longitudinal conductivity strongly increases below $T \lesssim \omega_c/4$ (this crossover scale, as well as the complete scaling function will be determined in Sec. 3), see also Fig. 1.

2.3.3. *Two-fluids regime:* $(l_B/\xi)v \ll T \ll v$

The peak longitudinal conductivity cannot diverge at vanishing temperature, and the law (7) must be cut-off by additional physical processes. Indeed, by further lowering the temperature, the Fermi distribution becomes sharp at the scale $T \ll v$, and the local Hall conductivity $\sigma_H(\mathbf{r})$ now assumes rapid spatial variations between quantized values $\nu e^2/h$ and $(\nu+1)e^2/h$, with ν the filling factor. The local conductivity model now reads

$$\sigma_H(\mathbf{r}) = \frac{e^2}{h}\nu + \frac{e^2}{h}\Theta[V(\mathbf{r}) + \mu - E_\nu], \quad (8)$$

introducing the step function Θ. The transport properties of this two-fluids model were considered extensively in previous works[20]. It was found using duality arguments that the local conductivity Eq. (8) leads in the (unphysical) limit of zero temperature to an exact value for the longitudinal peak conductivity in the $\sigma_0 \to 0$ limit: $\sigma_{xx}^{\text{peak}} = e^2/(2h)$. This result would seem at first sight at odds with the scaling Ansatz (2), which predicts a powerlaw vanishing of σ_{xx} at small σ_0. On mathematical grounds, the model Eq. (8) is quite peculiar in the sense that the fluctuations of the Hall conductivity $[\langle \sigma_H^2 \rangle - \langle \sigma_H \rangle^2]$ are actually diverging at the peak value, invalidating the Ansatz. Yet, the existence of a finite and universal value $\sigma_{xx}^{\text{peak}} = e^2/(2h)$ seems still physically surprising from the argumentation given in Sec. 2.2, where we argued that the fully opened current lines at σ_0 have a vanishing drift velocity at the saddle points of disorder. However, for such bimodal distribution of the local Hall conductivity Eq. (8) and in contrast to any continuous conductivity distribution, the drift velocity does not vanish anymore at the saddle points, which allows to establish a macroscopic current even in the absence of dissipation mechanisms. This simple argument allows to understand why the percolation scaling Ansatz (2) does not apply to the two-fluids model of Dykhne and Ruzin[20]. However, we will see below that other processes invalidate the model Eq. (8) in the limit of zero temperature. Moreover, we can also infer how the "exact" value $e^2/(2h)$ is approached from below. Indeed, the sharp Fermi function in Eq. (8) is always smeared on the scale T, recovering a continuous (but strongly non-Gaussian) distribution, leading likely to percolating-related power-law deviations from the exact zero-temperature result $e^2/(2h)$:

$$\sigma_{xx}^{\text{peak}} = \frac{e^2}{2h} - DT^\alpha \quad (9)$$

with a new critical exponent $\alpha > 0$ that is to our knowledge still unknown, and D some constant. The fact that the peak longitudinal conductivity levels off at low temperatures towards values close (but not strictly equal) to $e^2/(2h)$ has been noted from experimental data[20], see also Fig. 1.

2.3.4. *Wavefunction corrections:* $(l_B/\xi)^2 v \ll T \ll (l_B/\xi)v$

The low-temperature two-fluids conductivity model Eq. (8) relies on the high magnetic field limit, and is strictly speaking only correct in the limit $l_B \to 0$. However, quantum corrections will occur for finite l_B/ξ due to the fact that the electronic wavefunctions are not infinitely sharp transverse to the guiding center motion, but rather spread on the scale of the magnetic length l_B. For this reason, the local Hall conductivity $\sigma_{xx}(\mathbf{r})$ will not undergo infinitely sharp steps from a quantized value to the next as in Eq. (8), but rather rapid but smooth rises on the scale l_B, see Refs.17, 18. Because the wavefunctions extend transversely in a Gaussian manner, the resulting form of the Hall conductivity is easily understood (here for the lowest Landau level):

$$\sigma_H(\mathbf{r}) = \frac{e^2}{h} + \frac{e^2}{h} \int \frac{d^2 \mathbf{R}}{\pi l_B^2} \Theta \left[V(\mathbf{R}) - \mu - E_\nu\right] e^{-(\mathbf{r}-\mathbf{R})^2/l_B^2}. \tag{10}$$

Clearly, the two-fluid model Eq. (8) is recovered in the limit $l_B/\xi \to 0$, but for a more realistic smooth disorder, the correlation length ξ does not exceed a few hundreds of nanometers. In that case, the sharp step in Eq. (8) is smoothened whenever the new energy scale $(l_B/\xi)v$ sets in. Interestingly, we recover now a continuous conductivity distribution where the percolation Ansatz (2) should apply. Because the spatial fluctuations of the local Hall conductivity are no more controlled by temperature, we can infer without detailed calculation the following powerlaw for the peak longitudinal conductivity:

$$\sigma_{xx}^{\text{peak}} \propto T^{1-\kappa} \propto T^{3/13}. \tag{11}$$

Thus the peak conductivity should *decrease* again by cooling the sample to very low temperatures, as evidenced experimentally[8], see also Fig. 1.

2.3.5. *Onset of quantum tunneling:* $T \ll (l_B/\xi)^2 v$

By further cooling towards the limit of zero temperature, a new energy scale $(l_B/\xi)^2 v$ emerges, associated to quantum tunneling at the saddle points[18]. For high mobility samples, one can assume that transport remains incoherent between the widely separated saddle points, so that quantum interference effects can be neglected, and the local conductivity model Eq. (1) still applies (if not, non-local effects in the spirit of Ref. 21 must be accounted for). Here, the precise form of the local conductivity tensor is not yet fully understood, although a quasilocal approach that incorporates quantum tunneling can be developed[18]. For this reason, the precise scaling form of the peak longitudinal conductivity is still unknown in this regime, although a slower decrease than Eq. (11), leading to a kink at $T = (l_B/\xi)^2 v$, can be expected, due to the onset of the quantum processes allowing to transfer electrons above the saddle points:

$$\sigma_{xx}^{\text{peak}} \propto T^\beta \quad 0 < \beta < 3/13. \tag{12}$$

Such behavior was also observed experimentally in low temperature studies of the peak longitudinal conductivity[8], see also Fig. 1.

3. Diagrammatic approach to classical percolating transport

3.1. *Systematic weak coupling expansion and extrapolation to the percolation regime*

Our goal in this section is to discuss how classical percolation features in quantum Hall transport can be captured analytically by a diagrammatic approach[14], allowing to recover the percolation Ansatz (2) and accurate estimates of the critical exponent κ discussed in Sec. 2.3. Building on earlier works[22,23] for the case of the local conductivity tensor Eq. (1), one can show by standard techniques that the disorder averaged longitudinal macroscopic conductivity reads

$$\begin{pmatrix} \sigma_{xx} & -\sigma_{xy} \\ \sigma_{xy} & \sigma_{xx} \end{pmatrix} = \begin{pmatrix} \sigma_0 & -\langle \sigma_H \rangle \\ \langle \sigma_H \rangle & \sigma_0 \end{pmatrix} + \langle \hat{\chi}(\mathbf{r}) \rangle, \qquad (13)$$

where $\hat{\chi}(\mathbf{r})$ obeys the equation of motion:

$$\hat{\chi}(\mathbf{r}) = \delta\sigma(\mathbf{r})\hat{\epsilon} + \delta\sigma(\mathbf{r}) \int d^2 r' \, \hat{\epsilon} \, \hat{\mathcal{G}}_0(\mathbf{r}-\mathbf{r}') \hat{\chi}(\mathbf{r}'). \qquad (14)$$

We have introduced above the Hall conductivity fluctuations $\delta\sigma(\mathbf{r}) \equiv \sigma_H(\mathbf{r}) - \langle \sigma_H \rangle$, the antisymmetric 2×2 tensor $\hat{\epsilon}$, and the Green's function:

$$[\hat{\mathcal{G}}_0]_{ij}(\mathbf{r}) = \frac{\partial}{\partial r_i} \frac{\partial}{\partial r_j} \int \frac{d^2 p}{(2\pi)^2} \frac{e^{i\mathbf{p}\cdot\mathbf{r}}}{\sigma_0 |\mathbf{p}|^2 + 0^+}. \qquad (15)$$

In previous analyses of Eq. (14), several methods were proposed, such as a mean-field treatment[23], lowest order perturbation theory[24] in powers of $\langle [\delta\sigma]^2 \rangle / \sigma_0^2$, or self-consistent Born approximation[22]. Clearly these approaches are insufficient to capture the critical percolation behavior in the strong coupling limit $\sigma_0 \to 0$. However, the small dissipation Ansatz (2) ressembles the critical behavior typical of phase transitions, and leads hope that Padé extrapolation techniques of a sufficiently high order perturbative calculation could bridge the gap from weak (*i.e.* $\langle [\delta\sigma]^2 \rangle \ll \sigma_0^2$) to strong coupling (*i.e.* $\langle [\delta\sigma]^2 \rangle \gg \sigma_0^2$). The calculation actually simplifies for the case of Gaussian fluctuations of the local Hall conductivity $\delta\sigma(\mathbf{r})$, which applies to the highest temperature regimes considered in Eq. (3) and Eq. (5). By symmetry considerations, one finds that the Hall component is not affected in the high temperature regime, namely classical Hall's law $-\sigma_{xy} = en/B$ holds. By dimensional analysis, the longitudinal conductivity reads:

$$\sigma_{xx} = \sigma_0 + \sum_{n=1}^{\infty} a_n \frac{\langle \delta\sigma^2 \rangle^n}{\sigma_0^{2n-1}} \qquad (16)$$

with dimensionless coefficients a_n collecting all diagrams of order n in perturbation theory in $\langle \delta\sigma^2 \rangle / \sigma_0^2$. The longitudinal conductivity σ_{xx} thus receives non-trivial corrections that will lead to percolation effects in the limit $\sigma_0 \to 0$.

The methodology to compute the large σ_0 expansion relies in iterating Eq. (14) to the desired order, averaging over disorder owing to the relation (5), and evaluating the resulting multidimensional integral, either analytically or numerically, see Fig. 2. In order to simplify the calculations, we considered spatial correlations of disorder of the form $\langle \delta\sigma(\mathbf{r})\delta\sigma(\mathbf{r}') \rangle = \langle \delta\sigma^2 \rangle e^{-|\mathbf{r}-\mathbf{r}'|^2/\xi^2}$, with correlation length ξ, allowing us to compute the series Eq. (16) up to sixth loop order[14], see Table 1. Standard extrapolation techniques allow us to extract[14] the estimate $\kappa = 0.767 \pm 0.002$ for the critical exponent appearing in Eq. (2), quite close to the previously conjectured value[11,12,13,15] $\kappa = 10/13 \simeq 0.769$.

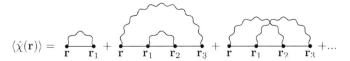

Fig. 2. Diagrammatic expansion in the case of Gaussian fluctuations of the local conductivity. Wiggly lines are associated to disorder averages, and solid lines to the Green's function Eq. (15).

Table 1. Coefficients a_n of the perturbative series Eq. (16) up to sixth loop order.

Order	Method	Coefficient a_n
1	Analytical	$\frac{1}{2}$
2	Analytical	$\frac{1}{8} - \frac{1}{2}\log(2)$
3	Analytical	0.2034560502
4	Numerical	-0.265 ± 0.001
5	Numerical	0.405 ± 0.001
6	Numerical	-0.694 ± 0.001

3.2. High temperature crossover function for $\sigma_{xx}^{\text{peak}}$

We finally provide a simple scaling function describing the crossover from the high temperature regime above the cyclotron energy $T \gg \hbar\omega_c$ to the intermediate situation $v \ll T \ll \hbar\omega_c$, where Gaussian fluctuations of the local Hall conductivity still arise, see Eq. (5). From this expression, we can connect the typical fluctuations of the Hall conductivity to the width $v = \sqrt{\langle [V(\mathbf{r})]^2 \rangle}$ of the disorder distribution: $\sqrt{\langle [\delta\sigma(\mathbf{r})]^2 \rangle} = \frac{e^2}{h} v \left| \sum_{m=0}^{\infty} n'_F(E_m - \mu) \right|$, so that the high temperature crossover function reads from the scaling Ansatz (2):

$$\sigma_{xx} = \sigma_0^{1-\kappa} \left| \frac{e^2}{h} v \sum_{m=0}^{\infty} n'_F(E_m - \mu) \right|^{\kappa}. \quad (17)$$

Note that for the Gaussian model studied here, a dimensionless prefactor in Eq. (17) happens[14] to be quite close to 1, and has not been written. In the regime $v \ll T \ll \hbar\omega_c$ and at the $\nu \to \nu + 1$ plateau transition, i.e. for $\mu = \hbar\omega_c(\nu + 1/2)$, we thus find:

$$\sigma_{xx}^{\text{peak}} = \sigma_0^{1-\kappa} \left| \frac{e^2}{h} \frac{v}{4T} \right|^{\kappa}, \quad (18)$$

recovering expression (7) for the longitudinal conductivity in the limit $v \ll T \ll \hbar\omega_c$.

We can alternatively re-express the sum over Landau levels in Eq. (17) by using Poisson summation formula[25] in the limit $T < \mu$, giving:

$$\sigma_{xx} = \sigma_0^{1-\kappa} \left[\frac{e^2}{h} \frac{v}{\hbar\omega_c} \left| 1 + \sum_{l=1}^{+\infty} (-1)^l \cos\left(\frac{2\pi l \mu}{\hbar\omega_c}\right) \frac{\frac{4\pi^2 l k_B T}{\hbar\omega_c}}{\sinh\left(\frac{2\pi^2 l k_B T}{\hbar\omega_c}\right)} \right| \right]^{\kappa} \quad (19)$$

vindicating expression (4) for the peak longitudinal conductivity in the $T \gg \hbar\omega_c$ limit. Either Eq. (17) or Eq. (19) can be used to extract the critical exponent κ from experimental data in the range of temperatures near the cyclotron energy.

4. Perspectives

As a conclusion, we list several issues that could be addressed in further developments of the present work.

- What is the magnetic field behavior of σ_{xx} at high temperature?
- Can one extract reliably the classical exponent κ from experiments?
- Are the exponents α and β of the low temperature regime related to κ?
- How do the finite probe currents affect the Hall plateau quantization?
- What are the fundamental differences between 2DEGs and graphene[26]?
- Is a more realistic description of electron-phonon conductivity[19] needed?
- Can one describe the crossover to Drude behavior at low magnetic fields[27]?
- Can one implement transport calculations using diagrammatic QMC[28]?

Acknowledgments

We thank S. Bera, A. Freyn, B. Piot, W. Poirier, M. E. Raikh, V. Renard and F. Schoepfer for stimulating discussions, and ANR METROGRAPH under Grant No. ANR-2011-NANO-004-006 for financial support.

References

1. K. Von Klitzing, G. Dorda, and M. Pepper, Phys. Rev. Lett. **45**, 494 (1980).
2. R. E. Prange and S. M. Girvin, *The Quantum Hall Effect*, (Springer, New York, 1987).
3. M. Janssen, O. Viehweger, U. Fastenrath, and J. Hadju, *Introduction to the Theory of the Integer Quantum Hall Effect* (VCH, Germany, 1994).
4. J. Matthews and M. E. Cage, J. Res. Natl. Inst. Stand. Technol. **110**, 497 (2005).
5. B. Huckestein, Rev. Mod. Phys. **67**, 357 (1995).
6. B. Kramer, T. Ohtsuki and S. Kettemann, Phys. Rep. **417**, 211 (2005).
7. F. Evers and A. D. Mirlin, Rev. Mod. Phys. **80**, 1355 (2008).
8. H. P. Wei *et al.*, Phys. Rev. B **45**, 3926 (1992).
9. W. Li *et al.*, Phys. Rev. B **81**, 033305 (2010).
10. K. Hashimoto *et al.*, Phys. Rev. Lett. **101**, 256802 (2008).
11. S. H. Simon and B. I. Halperin, Phys. Rev. Lett. **73**, 3278 (1994).
12. D. G. Polyakov and B. I. Shklovskii, Phys. Rev. Lett. **74**, 150 (1995).
13. M. M. Fogler and B. I. Shklovskii, Sol. State Comm. **94**, 503 (1995).
14. M. Flöser, S. Florens, and T. Champel, Phys. Rev. Lett. **107**, 176806 (2011).
15. M. B. Isichenko, Rev. Mod. Phys. **64**, 961 (1992).
16. M. R. Geller and G. Vignale, Phys. Rev. B **50**, 11714 (1994).
17. T. Champel, S. Florens and L. Canet, Phys. Rev. B **38**, 125302 (2008).
18. T. Champel and S. Florens, Phys. Rev. B **80**, 125322 (2009).
19. H. L. Zhao and S. Feng, Phys. Rev. Lett. **70**, 4134 (1993).
20. A. M. Dykhne and I. M. Ruzin, Phys. Rev. B **50**, 2369 (1994).
21. H. U. Baranger and A. D. Stone, Phys. Rev. B **40**, 8169 (1989).

22. Y. A. Dreizin and A. M. Dykhne, Sov. Phys. JETP **36**, 127 (1972).
23. D. Stroud, Phys. Rev. B **12**, 3368 (1975).
24. C. Timm, M. E. Raikh and F. von Oppen, Phys. Rev. Lett. **94**, 036602 (2005).
25. T. Champel and V. P. Mineev, Philos. Mag. B **81**, 55 (2001).
26. T. Champel and S. Florens, Phys. Rev. B **82**, 045421 (2010).
27. D. G. Polyakov, F. Evers, A. D. Mirlin and P. Wölfle, Phys. Rev. B **64**, 205306 (2001).
28. E. Gull *et al.*, Rev. Mod. Phys. **83**, 349 (2011).

FINITE SIZE SCALING OF THE CHALKER-CODDINGTON MODEL

KEITH SLEVIN

Department of Physics, Graduate School of Science, Osaka University
Machikaneyama 1-1, Toyonaka, Osaka 560-0043, Japan

TOMI OHTSUKI

Department of Physics, Sophia University
Kioi-cho 7-1, Chiyoda-ku, Tokyo 102-8554, Japan

In Ref. 1, we reported an estimate of the critical exponent for the divergence of the localization length at the quantum Hall transition that is significantly larger than those reported in the previous published work of other authors. In this paper, we update our finite size scaling analysis of the Chalker-Coddington model and suggest the origin of the previous underestimate by other authors. We also compare our results with the predictions of Lütken and Ross.[2]

Keywords: quantum Hall effect; Chalker-Coddington model; critical exponent.

PACS numbers: 73.43.-f, 71.30.+h

1. Introduction

When a strong magnetic field is applied perpendicular to an ideal two dimensional electron gas the kinetic energy of the electrons is quantized according to the formula $E_n = (n+1/2)\hbar\omega$. Here, n is a non-negative integer and ω is the cyclotron frequency $\omega = eB/m$. These Landau levels are highly degenerate and the density of states becomes a series of equally spaced delta functions. This degeneracy is broken by disorder and the Landau levels are broadened into Landau bands. Most of the electron states are Anderson localized with the exception of the states at the center of the Landau level where the localization length ξ has a power law divergence described by a critical exponent ν

$$\xi \sim |E - E_c|^{-\nu} . \qquad (1)$$

When the Fermi level is in a region of localized states, the Hall conductance is quantized in integer multiples of e^2/h. This effect is known as the quantum Hall effect.[3,4] Transitions between consecutive quantized values occur when the Fermi level passes through the center of a Landau band. This is a quantum phase transition. It is characterized by a two critical exponents. One is the critical exponent

ν mentioned above and the other is the dynamic exponent z, which describes temperature dependence.

The quantum Hall transition has been the subject of careful experimental study. The inverse of the product of the critical and dynamic exponents, $\kappa = 1/\nu z$, has been measured very precisely (see Table 1). The value of this product appears to be quite universal; the same value has been obtained in measurements in an $Al_x Ga_{1-x} As$ heterostructure[5] and in graphene.[6] The main problem is that an independent measurement of the dynamic exponent is needed to disentangle the values of the two exponents and, unfortunately, this has been measured much less precisely.

The quantum Hall transition has also been the subject of numerous numerical studies in models of non-interacting electrons. In earlier work, a consensus was reached (see Table 2) that $\nu \approx 2.4$ in apparent agreement with experiment. However, in 2009, we published[1] a numerical analysis of the Chalker-Coddington model in which we found a value of the exponent that was about 10% larger; a result which has since been confirmed by other authors (see Table 3). There now seems to be a consensus that the previous numerical work underestimated the exponent and the apparent agreement with experiment was a coincidence of errors. Below we discuss the reason for the previous underestimate of the exponent.

Another issue concerns the value of the dynamic exponent. For models of non-interacting electrons the dynamic exponent is known exactly, $z = 2$. However, this value cannot be compared directly with the experiment. Burmistrov et al.[17] have emphasized the distinction between the different dynamical exponents that occur in the problem. In the experiment of Li et al.[5] it seems clear that the dynamic exponent that is being measured describes the divergence of the phase coherence

Table 1. Experimental values of critical exponents for the quantum Hall transition.

	$\kappa = 1/z\nu$	ν	z
Experiment of Li et al.[5] ($Al_x Ga_{1-x} As$)	$0.42 \pm .01$	≈ 2.38	≈ 1
Experiment of Giesbers et al.[6] (graphene)	$0.41 \pm .04$	-	-

Table 2. Earlier estimates of the critical exponent ν.

Chalker and Coddington[7]	$2.5 \pm .5$	Huckestein and Kramer[8]	$2.34 \pm .04$
Mieck[9]	$2.3 \pm .08$	Huckestein[10]	$2.35 \pm .03$
Huo and Bhatt[11]	$2.4 \pm .1$	Lee and Wang[12]	$2.33 \pm .03$
Cain et al.[13]	$2.37 \pm .02$		

Table 3. Recent estimates of the critical exponent ν.

Slevin and Ohtsuki[1]	$2.593[2.587, 2.598]$	Obuse et al.[14]	$2.55 \pm .01$
Dahlhaus et al.[15]	$2.576 \pm .03$	Amado et al.[16]	$2.616 \pm .014$

length on approaching zero temperature

$$\ell_\varphi \sim T^{-z} \,. \tag{2}$$

It also seems reasonably safe to suppose that electron-electron interactions are the source of the electron dephasing. This does not mean however that electron-electron interaction are relevant in the renormalization group (RG) sense and that the quantum Hall transition is described by a fixed point in a theory of interacting electrons. A clear discussion of this can be found in Ref. 17. Also, as described in Ref. 18, it is thought that short range interactions are irrelevant in the RG sense and that only long range Coulomb interaction are relevant and would drive the system to a different interacting fixed point.

It has been pointed out to us by Alexei Tsvelik, that the value of the exponent we have found for the Chalker-Coddington model is very close to that predicted by Lütken and Ross. In a series of papers (Ref. 2 and references therein) these authors have argued that modular symmetry strongly constrains the possible critical theories of the quantum Hall transition. While we cannot claim to understand the details of the theory of Lütken and Ross, we attempt below to compare some of their key predictions with the results of our finite size scaling analysis.

2. Method

We calculated the Lyapunov exponents of the product of the transfer matrices for the Chalker-Coddington model.[7] This model describes electron localization in a two dimensional electron gas subject to a very strong perpendicular magnetic field. The basic assumption is that the random potential is smooth on the scale of the magnetic length. The electron wavefunctions are then concentrated on equipotentials of the random potential with tunneling between equipotentials at saddle points of the potential. In the Chalker-Coddington model this system is modeled by a network of nodes and links. A parameter x, which is essentially the energy of the electrons measured in units of the Landau band width relative to the center of the Landau band, fixes the tunneling probability at the nodes. A random phase distributed uniformly on $[0, 2\pi)$ is attached to each link to reflect the random length of the contours of the potential. For further details we refer the reader to the original article of Chalker and Coddington[7] and to the more recent review by Kramer et al.[19]

We considered the transfer matrix product associated with a quasi-one dimensional geometry with N nodes in the transverse direction and L nodes in the longitudinal direction.[a] The Lyapunov exponents of this random matrix product were estimated using the standard method.[20,21] The Lyapunov exponents are defined by taking the limit $L \to \infty$. By truncating the matrix product at a finite L, an estimate of the Lyapunov exponents was obtained. The sample to sample fluctuations

---

[a] For the detailed formulae see Ref. 1.

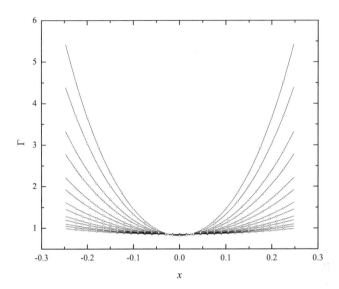

Fig. 1. The product Γ of the smallest positive Lyapnuov exponent γ and the number of nodes in the transverse direction $N = 4, 6, 8, 12, 16, 24, 32, 48, 64, 96, 128, 192, 256$ for the Chalker-Coddington model. The error in the data is much smaller than the symbol size. The lines are even order polynomial fits to the data for each N.

of this estimate decrease with the inverse of the square root of L. We performed a single simulation for each pair of x and N and truncated the transfer matrix product at a value of L that allowed estimation of the smallest positive Lyapunov exponent γ with a precision of 0.03%, except for the largest values of $N = 192$ and 256 where the precision was relaxed to either 0.05% or 0.1%. To ensure that simulations for different pairs of x and N were independent, we used the Mersenne Twister pseudo-random number generator MT2203 of Matsumoto et al.[22] provided in the Intel Math Kernel Library. All the simulations used a common seed. Independence was ensured by the use of a unique stream number for each simulation. We imposed periodic boundary conditions in the transverse direction for which choice the Lyapunov exponents are even functions of x. It is known that there is a critical point at the center of the Landau band, $x = 0$, and that, when the Fermi energy is driven through this point, the transition between Hall plateaux occurs.

To extract estimates of critical exponent and other quantities we used finite size scaling.[b] In this method, the behavior of the dimensionless quantity

$$\Gamma(x, N) = \gamma N, \qquad (3)$$

---

[b]This method was first applied to Anderson localization at about the same time by Pichard and Sarma[23,24] and by MacKinnon and Kramer[21,25]. See Ref. 26 for a pedagogical discussion.

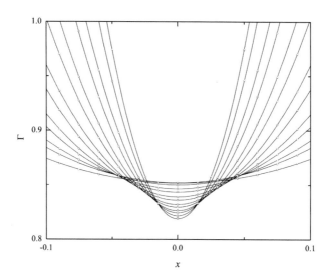

Fig. 2. The same data as in Fig. 1 but focussing on the critical point at $x = 0$. The residual variation of Γ with N at $x = 0$ is due to irrelevant scaling variables.

is analyzed as a function of both x and N. In the absence of any corrections to scaling we would expect this behavior to be described by the following finite size scaling law

$$\Gamma = F\left(N^{2\alpha} x^2\right), \qquad (4)$$

where F is an a priori unknown but universal scaling function and

$$\alpha = 1/\nu. \qquad (5)$$

Note that we have imposed the condition that Γ must be an even function of x.

The actual behavior of Γ as a function of x for different N is shown in Fig. 1 and, in more detail around $x = 0$, in Fig. 2. (The lines in the figures are polynomial fits. They will be discussed below.) According to Eq. (4) curves for different N should have a common crossing point at $x = 0$. However, it is clear from Fig. 2 that this is only approximately correct and that Γ is not exactly independent of N at $x = 0$ but varies by a several percent over the range of N studied. These corrections to scaling arise because of the presence of irrelevant scaling variables. These are variables with negative scaling exponents. Their effect is negligible for large N but their presence may lead to significant corrections at small N. This is consistent with what we see in Fig. 2.

To take account of corrections to scaling, we follow Ref. 27 and Ref. 28 and generalize the finite size scaling law

$$\Gamma = F\left(N^{2\alpha} v_0(x), N^{y_1} v_1(x), N^{y_2} v_2(x), \cdots\right). \tag{6}$$

In Eq. (6), v_0 is the relevant scaling variable and $v_1, v_2, \cdots$ are the irrelevant scaling variables. The associated exponents $y_1, y_2, \cdots$ are negative. The inclusion of irrelevant corrections permits the residual N dependence at $x = 0$ seen in Fig. 2 to be modeled. This form also allows for additional corrections due to non-linearities of the scaling variables as functions of x. To impose the condition that Γ must be an even function of x we restrict all the scaling variables to be even functions of x. In addition, since the critical point is at $x = 0$, we impose the condition that $v_0(0) = 0$. To fit the data, the function F is expanded as a Taylor series in all its arguments, and similarly the scaling variables. The coefficients in the Taylor series, together with the various exponents, play the role of fitting parameters.[c] The orders of truncation of the Taylor series are chosen sufficiently large to obtain an acceptable fit of the data (as measured using the χ^2-statistic and the goodness of fit probability). We have attempted this procedure with both one and two irrelevant corrections. Unfortunately, a stable fit of the data has eluded us. We have found that several fits of the data are possible. However, these fits do not yield mutually consistent estimates of the critical exponent.

3. Rudimentary finite size scaling

To circumvent the difficulties described in the previous section we resorted to a less sophisticated approach in which we abandoned the attempt to fit all the data in a single step. Instead, we fitted the data for each N independently to an even polynomial of x. For each N the order of the polynomial was chosen just large enough to give an acceptable goodness of fit. For $N = 4$, a quadratic was sufficient, while for $N = 256$ a sixth order polynomial was required. From these polynomials we estimated the curvature C of Γ at $x = 0$. The results are plotted in Fig. 3. The precision of the estimation of the curvature varies between 0.07% and 0.28%. According to (4) the curvature at $x = 0$ should vary with N as a power law,

$$C \equiv \left.\frac{d^2\Gamma}{dx^2}\right|_{x=0} \propto N^{2\alpha}. \tag{7}$$

This, of course, neglects corrections to scaling due to irrelevant variables and, indeed, a straight line fit to *all* the data does not yield an acceptable goodness of fit. However, if data for $N \leq 48$ are excluded, acceptable goodness of fits are obtained. The estimates of the critical exponent obtained in this way are tabulated in Table 4.

In Fig. 3 we have also plotted two lines. One is a solid line that corresponds to the straight line fit for $64 \leq N \leq 256$. The second is a dashed line with a

---

[c]Some extra conditions on the coefficients must be imposed to ensure that the model to be fitted to the data is unambiguous.

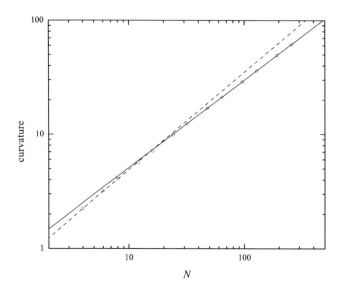

Fig. 3. The curvature C at $x = 0$ (see Eq. (7)), obtained from the polynomial fits shown in Figure 1, plotted as a function of N. The solid line is a straight line fit to the data for the largest five values of N. The slope corresponds to $\nu = 2.607$. The dashed line is a straight line with slope corresponding to $\nu = 2.34$ and passing though the $N = 4$ data point. The error in the data for the curvature is much smaller than the symbol size.

slope corresponding to the estimate of Huckestein[27] of the critical exponent and passing through the datum for the curvature at $N = 4$. While the main effect of the irrelevant corrections is the N dependent shift in the ordinate that is clearly visible in Fig.2, a smaller but not negligible effect on the curvature is also apparent in Fig.3. In our opinion, this is the reason why the critical exponent was underestimated in previous work. The precision of the numerical data was insufficient, and the range of N considered too small, for the irrelevant correction to be properly taken into account.

Table 4. Estimates of the critical exponent ν obtained from linear fits of $\ln C$ versus $\ln N$.

range of N considered	ν	Confidence intervals ($\pm 2\sigma$)
$64 \leq N \leq 256$	2.607	[2.598, 2.615]
$96 \leq N \leq 256$	2.599	[2.583, 2.616]
$128 \leq N \leq 256$	2.590	[2.586, 2.614]

4. Comparison with the predictions of Lütken and Ross

The predictions of the theory of Lütken and Ross that can be compared with the present work are for the critical exponent and the leading irrelevant exponent. Their theory contains a single unknown parameter, the central charge c. In terms of this parameter they predict that the critical exponent is[d]

$$\nu = \frac{10.42050633345819\cdots}{c}. \qquad (8)$$

In addition they predict that the leading irrelevant exponent and the critical exponent are related by

$$y = -\alpha = -1/\nu. \qquad (9)$$

While we are not aware of any justification for this, assuming $c = 4$ gives

$$\nu = 2.60512\cdots \qquad (10)$$

The most precise estimate in Table 4 is

$$\nu = 2.607 \pm 0.004. \qquad (11)$$

In fact, all of the estimates in Table 4 are consistent with the Lütken and Ross value. Turning to the irrelevant exponent, the situation is, unfortunately, much less clear. In Table 5 we tabulate some previous estimates of the irrelevant exponent. The Lütken and Ross prediction is

$$y = -0.383859\cdots \qquad (12)$$

The estimate of Huckestein[27] is consistent with this but subsequent estimates by Wang et al.[29] are somewhat ambiguous. In our opinion, further confirmation is needed before reaching a conclusion.

As mentioned above an unambiguous fit using Eq. (6) has not proved possible. The best we can do at present is to check the consistency of the Lütken and Ross values with our data by fixing both ν and y to these values when fitting. A series of such fits are tabulated in Table 6. In the fits only a single irrelevant variable is assumed and non-linearities in the scaling variables are ignored, i.e.

$$v_0 = v_{02} x^2, \quad v_1 = v_{10}. \qquad (13)$$

The scaling function is expanded as a Taylor series to order n_0 in the relevant field and order n_1 in the irrelevant field. One of the important quantities that can be

Table 5. Estimates from the literature of the leading irrelevant exponent y and the quantities used to estimate it.

Huckestein[27]	$y = -0.38 \pm .04$	Lyapunov exponents
Wang et al.[29]	$y \approx -0.52$	geometric average of the two-terminal conductance
Wang et al.[29]	$y \approx -0.72$	arithmetic average of the two-terminal conductance

[d]We noticed an error (a factor of 2) in Ref. 2. We thank Graham Ross for confirming this.

Table 6. Fits to the 379 data points shown in Figure 1.

n_0	n_1	number of parameters	χ^2	goodness of fit	Γ_c
3	3	16	398	0.10	$0.807 \pm .0005$
3	4	20	347	0.66	$0.804 \pm .0015$
3	5	24	343	0.67	$0.801 \pm .0035$

Table 7. Estimates of Γ_c obtained using published estimates of α_0 and Eq. (15)

	α_0	Γ_c
Obuse et al.[30]	2.2617 ± 0.0006	$0.8222 \pm .0019$
Evers et al.[31]	2.2596 ± 0.0004	$0.8156 \pm .0013$

estimated in this way is Γ_c, which is defined by

$$\Gamma_c = \lim_{N \to \infty} \Gamma(x=0, N) = F(0, 0, \ldots) . \qquad (14)$$

The quantity is significant because, if the quantum Hall critical theory has conformal symmetry, it is related to the multi-fractal exponent α_0 that occurs in the multi-fractal analysis of the wavefunction distribution at the critical point by

$$\Gamma_c = \pi(\alpha_0 - 2) . \qquad (15)$$

Some estimates of Γ_c obtained using published estimates of α_0 and Eq. (15) are tabulated in Table 7. Our numerical estimate of Γ_c is not completely consistent with those in the table. The reason for this is not yet clear.

5. Discussion

The agreement between our estimate for the critical exponent and prediction of Lütken and Ross is tantalizing but is it accidental, or does it have a deeper significance? A more precise numerical estimate of the irrelevant exponent is clearly highly desirable. In addition, the prediction of Lütken and Ross for the flow diagram in the $(\sigma_{xy}, \sigma_{xx})$ plane should also be amenable to numerical verification. We also need to know if a central charge $c = 4$ is physically justified.

Quite apart from whether or not the Lütken and Ross theory is exact for non-interacting electrons, the important question remains of clarifying the role of the electron-electron interactions in the observed critical behavior at the quantum Hall transition. More work along the lines of Ref. 32 might be very helpful in this regard.

Acknowledgments

We would like to thank Alexei Tsvelik for bringing the work of Lütken and Ross to our attention. This work was supported by Grant-in-Aid 23540376 and Korean WCU program Project No. R31-2008-000-10059-0.

References

1. K. Slevin and T. Ohtsuki, Physical Review B **80**, 041304(R) (2009).
2. C. A. Lutken and G. G. Ross, Physics Letters B **653**, 363 (2007).
3. K. v. Klitzing, G. Dorda, and M. Pepper, Physical Review Letters **45**, 494 (1980).
4. D. Yoshioka, *The quantum Hall effect, Springer series in solid-state sciences*, (Springer, Berlin ; New York, 2002).
5. W. Li et al., Physical Review Letters **102**, 216801 (2009).
6. A. J. M. Giesbers et al., Physical Review B **80**, 241411 (2009).
7. J. T. Chalker and P. D. Coddington, Journal of Physics C: Solid State Physics **21**, 2665 (1988).
8. B. Huckestein and B. Kramer, Physical Review Letters **64**, 1437 (1990).
9. B. Mieck, EPL (Europhysics Letters) **13**, 453 (1990).
10. B. Huckestein, EPL (Europhysics Letters) **20**, 451 (1992).
11. Y. Huo and R. N. Bhatt, Physical Review Letters **68**, 1375 (1992).
12. D.-H. Lee and Z. Wang, Philosophical Magazine Letters **73**, 145 (1996).
13. P. Cain, R. A. Romer, and M. E. Raikh, Physical Review B **67**, 075307 (2003).
14. H. Obuse et al., Physical Review B **82**, 035309 (2010).
15. J. P. Dahlhaus, J. M. Edge, J. Tworzydlo, and C. W. J. Beenakker, Physical Review B **84**, 115133 (2011).
16. M. Amado et al., Physical Review Letters **107**, 066402 (2011).
17. I. S. Burmistrov et al., Annals of Physics **326**, 1457 (2011).
18. A. M. M. Pruisken, International Journal of Modern Physics B **24**, 1895 (2010).
19. B. Kramer, T. Ohtsuki, and S. Kettemann, Physics Reports **417**, 211 (2005).
20. I. Shimada and T. Nagashima, Progress of Theoretical Physics **61**, 1605 (1979).
21. A. MacKinnon and B. Kramer, Zeitschrift für Physik B Condensed Matter **53**, 1 (1983).
22. M. Matsumoto and T. Nishimura, in *Monte Carlo and Quasi-Monte Carlo Methods 1998*, edited by H. Niederreiter and J. Spanier (Springer, Berlin ; New York, 1999), pp. 56–69.
23. J. L. Pichard and G. Sarma, Journal of Physics C: Solid State Physics L127 (1981).
24. J. L. Pichard and G. Sarma, Journal of Physics C: Solid State Physics L617 (1981).
25. A. MacKinnon and B. Kramer, Physical Review Letters **47**, 1546 (1981).
26. D. J. Amit and V. Martin-Mayor, *Field theory, the renormalization group, and critical phenomena*, 3rd ed. ed. (World Scientific, New Jersey ; London, 2005).
27. B. Huckestein, Physical Review Letters **72**, 1080 (1994).
28. K. Slevin and T. Ohtsuki, Physical Review Letters **82**, 382 (1999).
29. X. Wang, Q. Li, and C. M. Soukoulis, Physical Review B **58**, 3576 (1998).
30. H. Obuse et al., Physical Review Letters **101**, 116802 (2008).
31. F. Evers, A. Mildenberger, and A. D. Mirlin, Physical Review Letters **101**, 116803 (2008).
32. B. Huckestein and M. Backhaus, Physical Review Letters **82**, 5100 (1999).

Localisation 2011
International Journal of Modern Physics: Conference Series
Vol. 11 (2012) 70–78
© World Scientific Publishing Company
DOI: 10.1142/S2010194512006174

BULK AND EDGE QUASIHOLE TUNNELING AMPLITUDES IN THE LAUGHLIN STATE

ZI-XIANG HU

Department of Electrical Engineering, Princeton University, Princeton, New Jersey 08544, USA

KI HOON LEE

Asia Pacific Center for Theoretical Physics, Pohang and Department of Physics, Pohang University of Science and Technology, Pohang, Gyeongbuk 790-784, Korea

XIN WAN

Zhejiang Institute of Modern Physics, Zhejiang University, Hangzhou 310027, P. R. China

The tunneling between the Laughlin state and its quasihole excitations are studied by using the Jack polynomial. We find a uniyersal analytical formula for the tunneling amplitude, which can describe both bulk and edge quasihole excitations. The asymptotic behavior of the tunneling amplitude reveals the difference and the crossover between bulk and edge states. The effects of the realistic coulomb interaction with a background-charge confinement potential and disorder are also discussed. The stability of the tunneling amplitude manifests the topological nature of fractional quantum Hall liquids.

Keywords: Laughlin state, quasiparticle tunneling, Jack polynomial, scaling, disorder

PACS numbers: 73.43.Cd, 73.43.Jn

1. Introduction

Fractional quantum Hall (FQH) liquids are examples of experimentally realizable phases that support topological objects. Quasiparticle excitations in the FQH liquids have fractional charge and obeys fractional statistics.[1,2,3] Some FQH liquids may support more exotic quasiparticle excitations with non-Abelian statistics, which have potential applications in topological quantum computation.[4] The measurement of the transport properties of the quasiparticles propagating along the edge of FQH liquids is crucial for the identification of the topological nature of the systems. As standard practice in noise and interference experiments quantum point contacts are introduced to allow quasiparticles propagating on one edge to tunnel to another. This motivated the authors and their collaborators to study the quasiparticle tunneling amplitudes in FQH liquids in the disk geometry.[5,6] We found that the tunneling amplitudes exhibit interesting scaling behavior, whose exponent is related to the conformal dimension of the tunneling quasiparticles.

In the disk geometry with an open boundary edge excitations arise from the bosonic density deformation of the FQH liquids and, in the case of the Moore-Read state, from an extra branch of Majorana fermion mode. The edge excitations are closely related to the bulk quasihole excitations. For example, a charge $|e|/m$ quasihole at ξ in a $\nu = 1/m$ Laughlin droplet $\Psi_L = \prod_{i<j}(z_i - z_j)^m \exp(-\sum_i |z_i|^2/4)$ of N electrons can be described by the wavefunction

$$\Psi_{qh} = \left[\prod_i (z_i - \xi)\right] \Psi_L, \qquad (1)$$

where $z_j = x_j + iy_j$ is the complex coordinate for the jth electron. The quasihole excitation can be expanded into a sum of edge excitations, whose amplitudes depend on the location of the quasihole ξ

$$\Psi_{qh} = \left[\sum_n (-\xi)^{N-n} s_n\right] \Psi_L, \qquad (2)$$

where $s_n = \mathcal{S}_N(\prod_i^n z_i)$ is a symmetric polynomial of degree n and $\mathcal{S}_N$ denotes the total symmetrization among the N coordinates. The first few examples are $s_0 = 1$, $s_1 = \sum_i z_i$, $s_2 = \sum_{i<j} z_i z_j$, etc. In a Laughlin state the gapless edge mode $s_n \Psi_L$ spans the Hilbert space of low-energy edge excitations. In fact, there is no strict distinction between a quasihole and an edge excitation with a large angular momentum ΔM. The conventional understanding is that an edge excitation has $\Delta M = O(1)$, while a quasihole excitation $\Delta M = O(N)$. The correspondence of the bulk and edge excitations suggests that the bulk and edge quasihole tunneling amplitudes may have a parallel correspondence and, therefore, a smooth crossover.

In this paper we confirm that the quasihole tunneling amplitude between the Laughlin state and its bulk and edge excitations ($\langle \Psi_{qh}|\mathcal{T}|\Psi_L\rangle$ and $\langle s_n \Psi_L|\mathcal{T}|\Psi_L\rangle$, where $\mathcal{T}$ is the tunneling operator to be defined below), can be described by a unified picture. In particular, we conjecture a universal formula in the limit of a small interedge distance, which can be reduced to the bulk quasihole tunneling amplitude result reported earlier.[6] The tunneling amplitude of topological nature is robust against the influence of long-range interaction and disorder. The paper is organized as follows. In Sec. 2, we review our previous study about the bulk quasihole tunneling amplitude in Laughlin state and explain the technical details. The study is extended to edge quasihole tunneling amplitude in Sec. 3. The robustness of the tunneling amplitude in the presence of long-range coulomb interaction and disorder is emphasized in Sec. 4. We summarize the results in Sec. 5.

2. A Brief Review of the Bulk Quasihole Tunneling amplitude in a Laughlin Droplet

We consider a FQH liquid on a disk and assume a single-particle potential $V_{\text{tunnel}}(\theta) = V_t \delta(\theta)$, which breaks the rotational symmetry. The potential defines a tunneling path for quasiparticles under the gate influence at a quantum point contact. We start by quoting the result[5] for the tunneling matrix element between two single-particle states with angular momentum k and l, $v_p(k,l) \equiv \langle k|V_{tunnel}(\theta)|l\rangle = \frac{V_t}{2\pi}\frac{\Gamma((k+l)/2+1)}{\sqrt{k!l!}}$. The tunneling operator of the many-body wavefunction is then defined as $\mathcal{T} = \sum_i V_{tunnel}(\theta_i)$ and the amplitude for a quasihole to tunnel to the droplet edge is

$$\Gamma_{qh} = \frac{\langle\Psi_{qh}|\mathcal{T}|\Psi_0\rangle}{\sqrt{\langle\Psi_{qh}|\Psi_{qh}\rangle}\sqrt{\langle\Psi_0|\Psi_0\rangle}}, \quad (3)$$

where $|\Psi_0\rangle$ and $|\Psi_{qh}\rangle$ are the wavefunctions for the ground state and the quasihole state, respectively.

The matrix element form of the tunneling amplitude suggests that we can either use Lanczos-type exact diagonalization or variational Monte Carlo simulation to calculate. However, these approximations fail to reach the accuracy needed for the error-free determination of the conformal dimensions of quasiholes (though they are useful in the discussion of the long-range interaction and disorder effects). Fortunately, the application of Jack polynomial provides an instructive yet numerical exact calculation method.

Let us digress and explain first the connection between the Laughlin model wavefunction, which is the exact ground state of the hard-core (V_1 only) interaction, and the Jack polynomial.[7] In general, Jacks belong to a family of *symmetric* multivariate polynomials of the complex particle coordinates. Potentially, they can be FQH wavefunctions for bosons (appending the ubiquitous Gaussian factor) or for fermions (with an extra antisymmetric factor $\prod_{i<j}(z_i - z_j)$, i.e., the Vandermonde determinant). A Jack $J^\alpha_\lambda(z_1, z_2, \cdots, z_N)$ can be parametrized by a rational number α (negative in this context), which is related to the clustering properties of the polynomial wavefunction, and a root configuration λ, which satisfies a generalized Pauli exclusion principle and from which one can derive a set of monomials that form a basis for the Jack. The Jack is an eigenstate of the corresponding Calogero-Sutherland Hamiltonian

$$H_{CS} = \sum_i (z_i\partial_i)^2 + \frac{1}{\alpha}\sum_{i<j}\frac{z_i+z_j}{z_i-z_j}(z_i\partial_i - z_j\partial_j), \quad (4)$$

where $\partial_i \equiv \partial/\partial z_i$. Take the bosonic Laughlin state at $\nu = 1/2$ (which corresponds to the fermionic Laughlin state at $\nu = 1/3$) for a concrete example. One can easily check that $\prod_{i<j}(z_i - z_j)^2$ satisfies H_{CS} with $\alpha = -2$, which is related to the fact that the bosonic (or the corresponding fermionic) wavefunction vanishes as $(z_i - z_i)^2$ [or $(z_i - z_j)^3$] when particle i approaches j. For two bosons, one obviously has

$$(z_1 - z_2)^2 = z_1^2 z_2^0 - 2z_1^1 z_2^1 + z_1^0 z_2^2 = 1\cdot\mathcal{S}_2(z_1^2 z_2^0) + (-2)\cdot\mathcal{S}_2(z_1^1 z_2^1), \quad (5)$$

which is an expansion of the polynomial wavefunction into a sum of symmetric monomials. The N-particle wavefunction $\prod_{i<j}(z_i - z_j)^2$ can be expanded as

$$1\cdot\mathcal{S}_N(z_1^{2N-2} z_2^{2N-4}\cdots z_N^0) + (-2)\cdot\mathcal{S}_N(z_1^{2N-3} z_2^{2N-3} z_3^{2N-6} z_4^{2N-8}\cdots z_N^0) + \cdots. \quad (6)$$

This crude example (perhaps with a more elaborate expansion) illustrates the idea that a bosonic (fermionic) wavefunction can be expanded by a set of homogeneous symmetric monomials (Slater determinants), which can be derived from a single monomial, known as the root, using a squeezing rule, which lowers the relative angular momentum and conserves the total angular momentum for the system with rotational invariance. We choose a numeric string representation for the root configuration, which is simply the collection of occupation numbers of the single-particle orbitals (z^m in the quantum Hall context). The root configuration for the above example of N-particle bosonic Laughlin state is, therefore, $1010\cdots 101$ and for the corresponding fermioic one $100100\cdots 1001$. Let us use the convention that the leftmost digit corresponds to the z^0 orbital, i.e., the droplet center. It is closely related to the topological nature of the wavefunction that the mere knowledge of a Jack parameter and a matching root configuration are enough to generate the coefficients of all the descendant symmetric monomials (Slater determinant) numerically *exact* – practically to more than 10 particles – with a recursive method.[8]

The FQH model wavefunctions can also be written as the correlators of certain primary fields in some conformal field theories. For example, the Laughlin wavefunction at filling fraction $\nu = 1/M$ can be constructed by the chiral boson conformal field theory (CFT) with a compactification radius M.[9] The primary fields are vertex operators $e^{im\varphi(z)/\sqrt{M}}$, where $\varphi(z)$ is a chiral boson field. Operators with $m = 1, 2, \ldots M - 1$ correspond to quasiholes with different charge, whose corresponding conformal dimensions are $\Delta = m^2/(2M)$. It is, therefore, reasonable to expect that the tunneling amplitude Γ as a function of system size N (with a given tunneling distance d) may show scaling behavior, $\Gamma \sim N^\beta$, whose scaling exponent β is related to the conformal dimension of the tunneling particle. Our previous work[6] confirmed the hidden (due to the dominant single-particle effect) scaling behavior of the quasihole tunneling amplitude, $\beta = 1 - 2\Delta$, which can be explained by the effective field theory consideration that the tunneling amplitudes contains factors of quasiparticles propagating along the opposite edges.

In the identification of the relation between the scaling exponent and the conformal dimension of the corresponding quasiparticle, the exact calculation of the tunneling amplitude $\Gamma = \langle \Psi_{qh}|\mathcal{T}|\Psi_L\rangle / (\sqrt{\langle \Psi_{qh}|\Psi_{qh}\rangle}\sqrt{\langle \Psi_L|\Psi_L\rangle})$ using the Jack polynomial method plays a crucial role, which, in fact, motivated the effective field theory interpretation.[6] The exactness allowed the elimination of all finite-size uncertainties with the conjecture of an exact tunneling amplitude at the small interedge limit, at which we deform the N-particle system (with a puncture, or quasiholes, at the center) into a ribbon, or topologically $S^1 \times [0,d]$ ($d/l_B \ll N$). This effectively erases the dominant single-particle effect (i.e. the Gaussian Landau level form factor). In this limit, for $M = 3$ or $\nu = 1/3$, we conjectured[6] that the tunneling amplitude for the charge-$e/3$ quasihole is

$$2\pi \Gamma_{L,M}^{e/M}(N) = \frac{N}{M} B\left(N, \frac{1}{M}\right), \tag{7}$$

where $M = 3$ and N is the number of electrons. Here we introduce the beta function $B(x, \eta) = \Gamma(x)\Gamma(\eta)/\Gamma(x + \eta)$ which, for large x and fixed η, asymptotically approaches $\Gamma(\eta)x^{-\eta}$, where $\Gamma(x)$ is the Gamma function (not the tunneling amplitude elsewhere). We verified numerically that the conjecture is *exact for up to 10 electrons*; therefore, asserting the conjecture is also exact for larger system, we obtain the exact exponent $\beta^{(e/3)} = 1 - 1/3 = 2/3$ in the scaling behavior $\Gamma^{(e/3)} \sim N^{\beta^{(e/3)}}$. This is also verified to be applicable for $M = 5$.[a] In other words, based on the scaling analysis we discussed earlier, we can compute the conformal dimension of smallest charged quasiholes in the $\nu = 1/M$ Laughlin state to be $\Delta = 1/(2M)$.

The quasi-one-dimensional ribbon limit essentially removes the unnecessary geometrical information of the wavefunctions, and subsequently reveals a perfect scaling behavior otherwise embedded in inaccuracy and deviations due to small system size. A similar consideration, dubbed as the conformal limit, allows the opening of a full gap in the entanglement spectra of systems on sphere geometry,[10] in a way that the low-lying levels showing the universal conformal field theory counting are well separated from the higher Coulomb ones, which are not universal. In both cases, we found that topology stands out after we suppress the geometrical information. In other words, the topological information is encoded in the set of coefficients in Eq. (6), while the geometrical information is encoded in the monomials, which may be deformed, for example, to accommodate the geometry of shear transformation and rotation.[11]

3. The Edge Quasihole Tunneling Amplitude in a Laughlin Droplet

In the last section, we consider the tunneling amplitude of the bulk quasihole at the center of the droplet. In fact, we can generalize the discussion to quasiholes located elsewhere, or to edge excitations. The generalization is straightforward for the Jack polynomial approach, which applies to low-lying excitations, such as the edge mode in disk geometry[12] and the magnetic-roton mode in sphere geometry.[13] The study of the tunneling properties of edge states ($s_n \Psi_L$, $n = 1, 2, 3, \ldots$) is an important tool for understanding the topological bulk states both from experimental and theoretical point of view. Here we focus on the tunneling amplitude $\Gamma = \langle s_n \Psi_L | \mathcal{T} | \Psi_L \rangle$, which crossover smoothly from the tunneling of a bulk quasihole excitation to edge excitations, as n decreases from N to 1.

The edge mode for the Laughlin state corresponds to a set of states whose root configuration are: ...10010010001, ...10010001001, ...10001001001, etc., meaning one 0 (or a quasihole) can be inserted in any one of the 100 unit cells. The tunneling problem we study here is the amplitude for the quasihole in these edge states to tunnel to the outer edge, leaving the Laughlin state behind. Again, we focus on the quasi-one-dimensional ribbon limit to look for a unified analytical solution. As we have already demonstrated the exactness of the method elsewhere,[6] we focus here

---

[a] Eq. (7) also applies to the integer case ($M = 1$), in which the righthand side reduces to unity.

on presenting the analytical conjecture on the tunneling amplitude, which has been verified to be correct for all accessible system sizes.

For the consistency with Eq. (7), we define,

$$T(N) \equiv T(N, 1/M) = \sqrt{\frac{2\pi}{M}\Gamma^{e/M}_{L,M}(N)} = \frac{1}{M}\sqrt{NB\left(N, \frac{1}{M}\right)}, \quad (8)$$

for $N \geq 0$ and specifically define $T(0, 1/M) = 1/M$. This allows us to unify the edge and bulk quasihole tunneling amplitudes as

$$2\pi\Gamma^{e/M}_{L,M}(N, \Delta k) = \frac{T(N)T(\Delta k)}{T(N - \Delta k)}, \quad (9)$$

where the integer Δk is the angular momentum of the edge/quasihole excitation (i.e., the number of 1s to the right of the inserted 0). For example, if we consider a system with a root configuration 10010010010001001 and a Jack parameter $\alpha = -2$, the additional parameters can be read as $N = 6$, $M = 3$, and $\Delta k = 2$. When $\Delta k = N$, it recovers Eq. (7), as the additional 0 is located at the leftmost position. Note when $\Delta k = 0$, the tunneling amplitude measures the average density, which is $1/M$. In the thermodynamic limit, for edge excitations, i.e., $\Delta k = O(1)$, we find

$$2\pi\Gamma^{e/M}_{L,M}(N, \Delta k) = T(\Delta k) + O(1/N). \quad (10)$$

For bulk excitations, i.e., $\Delta k = O(N)$, we find

$$2\pi\Gamma^{e/M}_{L,M}(N, \Delta k) \sim N^{1-1/M}. \quad (11)$$

These results allow us to compute straightforwardly the tunneling amplitude for a quasihole anywhere inside a Laughlin droplet to the edge.

4. Robustness of Tunneling Amplitude in the Presence of Realistic Edge Confinement and Disorder

So far we discussed the exact tunneling results using model wavefunctions generated as Jack polynomials. In other words, the Laughlin wavefunction and its edge states (including the single-quasihole state) are the eigenfunction of the Hamiltonian for electrons with hard-core interaction. In a realistic GaAs/GaAlAs heterostructure, the electrons interact with each other via a long-range coulomb repulsion and are confined by a neutralizing background charge confinement from a doping layer at a setback distance d. To see whether the tunneling amplitude is robust, we need to verify the validity of Eq. (7) in the presence of Coulomb interaction. Based on a previous study,[14] we fix the setback distance of the background charge at $d = 0.5\ l_B$, at which the system is in the Laughlin phase, i.e., the global ground state has the same quantum number as that of the Laughlin state and a close-to-unity overlap with the latter as well. On the other hand, the quasihole state is produced by a Gaussian impurity potential $H_W = W_g \sum_m \exp(-m^2/2s^2) c^+_m c_m$ with a width $s = 2\ l_B$,[15] which models, e.g., the STM tip potential in an experiment. We calculate the bulk quasihole tunneling amplitude up to 10 electrons and compare

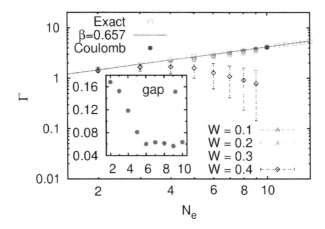

Fig. 1. The tunneling amplitude for the Laughlin phase in the presence of long-range Coulomb interaction and disorder in the quasi-one-dimensional ribbon limit. The exact results for the Laughlin state [Eq. (7)] can be fit by a power law with an exponent 0.657 (the exact value should be 2/3, which is prone to finite-size error in fitting). In the realistic model with Coulomb interaction and neutralizing charge confinement at a setback distance $d = 0.5 \, l_B$ (solid points), we obtain almost the same Γ as using the model wavefunctions. When we include the disorder potential with strength W, Γ remains unchanged at weak disorder, but deviates from the exact values at strong disorder. The inset plots the energy gap between the ground state and the first excited state in the same angular momentum subspace with $M = 3N(N-1)/2$ of the pure Coulomb Hamiltonian with $d = 0.5 \, l_B$.

it (deformed to the quasi-one-dimensional ribbon limit) with Eq. (7). As shown in Fig. 1, the long-range coulomb interaction has very little effect on the tunneling amplitude, hence the scaling behavior of Eq. (7) is, to a good approximation, valid for the realistic interaction, as long as the system remains in the Laughlin phase.

In addition, a realistic system also contains impurity scatterings. Nevertheless, the topological properties of an FQH state is believed to be robust against weak disorder. To prove the statement regarding disorder, we consider an uncorrelated random potential on each Landau level orbital, such that $H_D = \sum_m U_m c_m^+ c_m$, where U_m denotes the random potential on the mth orbital, whose value is randomly chosen in the range of $[-W/2, W/2]$. We compute the tunneling amplitudes by averaging over more than 1000 random samples for a given disorder strength W. As shown in Fig. 1, the tunneling amplitudes for weak disorder, i.e., $W = 0.1$, are almost the same as that in the pure coulomb case for all accessible system sizes. However, when we increase the strength of disorder gradually, the tunneling amplitude Γ deviates significantly from the exact results for large enough system size and therefore, the scaling hypothesis of Eq. (7) fails in the strong disorder case. To quantitatively understand the disorder effect, we can define and compute the energy gap for the system as the energy difference between the ground state (which is the Laughlin-like state) and the first excited state in the same subspace with a total angular momentum $3N(N-1)/2$. As shown in the inset of Fig. 1, when we fix $d = 0.5\ l_B$, the energy gap has very little finite size fluctuations for $N > 6$; the gap is around $0.06\ e^2/\epsilon l_B$. Therefore, disorder starts to affect the tunneling amplitudes when the strength of disorder is significantly larger than the energy gap (up to an $O(1)$ prefactor). This, in return, suggests that the deviation of the tunneling amplitude from the scaling behavior can be explored to study the transition of the FQH phase to insulator, much like the Chern number study of the FQH-insulator transition.[16]

5. Conclusions

In summary, we calculate the tunneling amplitude for quasihole in the Laughlin phase, generalizing the previous result for the tunneling of a quasihole at the center of a circular Laughlin droplet to an arbitrary location. This is achieved by considering the tunneling amplitudes between the Laughlin state and its accompanying edge states. Using the exact Jack polynomial expansion, we showed that the bulk and edge quasihole tunneling can be unified by a single equation (9) for any system size. We also demonstrate that the quasihole tunneling amplitude is robust against realistic considerations, such as the long-range coulomb interaction, neutralizing background charge confinement, and moderate amount of disorder.

Acknowledgments

This work was supported by the National Basic Research Program of China (973 Program) grant No. 2009CB929100, National Natural Science Foundation of China (NSFC) grant No. 11174246, and US DOE grant No. DE-SC0002140. KHL acknowledges the support at the Asia Pacific Center for Theoretical Physics from the Max Planck Society and the Korean Ministry of Education, Science and Technology.

References

1. F. Wilczek, Phys. Rev. Lett. **48**, 1144 (1982).
2. B. I. Halperin, Phys. Rev. Lett. **52**, 1583 (1984).
3. D. Arovas, J. R. Schrieffer, and F. Wilczek, Phys. Rev. Lett. **53**, 722 (1984).
4. A. Kitaev, Ann. Phys. **303**, 2 (2003).
5. H. Chen, Z.-X. Hu, K. Yang, E. H. Rezayi, and X. Wan, Phys. Rev. B **80**, 235305 (2009).
6. Z.-X. Hu, K.-H. Lee. E. H. Rezayi, X. Wan and K. Yang, New J. Phys. **13**, 035020 (2011).
7. B. A. Bernevig and F. D. M. Haldane, Phys. Rev. Lett. **100**, 246802 (2008). B. Bernevig and F. Haldane, Phy. Rev. Lett. **101**, 246806 (2008). B. A. Bernevig and N. Regnault, Phys. Rev. Lett. **103**, 206801 (2009).
8. R. Thomale, B. Estienne, N. Regnault, and B. A. Bernevig, Phys. Rev. B **84**, 045127 (2011).
9. G. Moore and N. Read, Nucl. Phys. B **360**, 362 (1991).
10. R. Thomale, A. Sterdyniak, N. Regnault, and B. A. Bernevig, Phys. Rev. Lett. **104**, 180502 (2010).

11. N. Read and E. H. Rezayi, Phys. Rev. B **84**, 085316 (2011).
12. K.-H. Lee, Z.-X. Hu, and X. Wan (unpublished).
13. B. Yang, Z.-X. Hu, Z. Papić, and F. D. M. Haldane, arXiv:1201.4165 (unpublished).
14. X. Wan, K. Yang, and E. H. Rezayi, Phys. Rev. Lett. **88**, 056802 (2002).
15. Z.-X. Hu, X. Wan, and P. Schmitteckert, Phys. Rev. B **77**, 075331 (2008).
16. D. N. Sheng, X. Wan, E. H. Rezayi, K. Yang, R. N. Bhatt, and F. D. M. Haldane, Phys. Rev. Lett. **90**, 256802 (2003).

"RARE" FLUCTUATION EFFECTS IN THE ANDERSON MODEL OF LOCALIZATION

R.N. BHATT

*Department of Electrical Engineering and Princeton Center for Theoretical Science,
Princeton University Princeton, NJ 08544, USA
ravin@princeton.edu*

S. JOHRI

*Department of Electrical Engineering, Princeton University
Princeton, NJ 08544, USA
sjohri@princeton.edu*

Received 19 March 2012

We discuss the role of rare fluctuation effects in quantum condensed matter systems. In particular, we present recent numerical results of the effect of resonant states in Anderson's original model of electron localization. We find that such resonances give rise to anomalous behavior of eigenstates not just far in the Lifshitz tail, but rather for a substantial fraction of eigenstates, especially for intermediate disorder. The anomalous behavior includes non-analyticity in various properties as a characteristic. The effect of dimensionality on the singularity, which is present in all dimensions, is described, and the behavior for bounded and unbounded disorder is contrasted.

1. Introduction: A Brief History of Large Disorder and Rare Fluctuations

Rare fluctuation phenomena received little attention in the early days of Solid State Physics. Ignoring them seems to have been the norm – what effects could such low probability phenomena possibly have, especially for a thermodynamically large system? On the contrary, the phenomenal success of Bloch's theorem and its application to wave phenomena in solids (e.g. electron energy bands, phonon dispersions), led to a kind of "Blochitis" among the practitioners of the field. Materials not having proper translational symmetry (i.e. disordered materials like amorphous solids, glasses, liquids) were addressed using perturbative techniques starting from the uniform system, usually a crystal with an appropriate coordination number, and then averaging the variables representing the disorder. This led to the growth of various mean-field like schemes for treating disorder, like the Coherent Potential Approximation, the averaged T-matrix approximation, the Effective Medium Approximation etc.

However, none of these methods quite captured effects that were primarily present because of disorder, as exemplified in the seminal paper by Anderson[1] in 1958 entitled "Absence of Diffusion in Certain Random Lattices". Anderson showed, for the first time, the existence of localized states in a disordered one-electron system, localized around disorder-specific locations in the system, in contrast to the Bloch waves in crystalline materials, which extended in a periodic fashion over the whole macroscopic system at all allowed energies. In our three dimensional world, localized states existed for moderate to large disorder. Further, as emphasized by Mott[2], these localized states were separated in energy from extended states; the separatrix was called the mobility edge; in recent years, it is also referred to as the critical energy of the localization-delocalization transition, E_C. Thus, with increasing disorder, localized states appeared first in regions of low density of states (band tails), and gradually the mobility edge(s) moved towards the center of the band, until at a critical disorder, all states became localized and no extended Bloch-like states were left. (In reality, extended states for moderate disorder also exhibit large-scale fluctuations that are disorder specific, and are known in the field as mesoscopic fluctuations).

Soon after Anderson's paper, Mott and Twose[3] showed that all electronic states were localized in one dimension for arbitrarily weak disorder (and hence Bloch's ideas broke down immediately upon introduction of disorder, at least as far as electron transport was concerned). It took almost two decades, using renormalization group (RG) ideas, to establish that in two dimensions[4] all states were again localized with arbitrary small disorder in the case of pure potential scattering, like the one considered by Anderson in his 1958 paper. In hindsight, given the situation in lower dimensions, it seems lucky that we live in a three-dimensional world, where many of the properties of crystalline materials we love and use are robust, at least to small amounts of imperfections.

Following Anderson's stylized model, Lifshitz[5] considered a somewhat more realistic model for problem of impurity bands in doped semiconductors, in which positional randomness of the dopants was the dominant source of disorder. Here the dopant atom, often less than one part in a million, dominates all the characteristics of the material that give it the unique properties that led to the semiconductor revolution. It is ironic to note that while solid state physicists studied doped semiconductors quite extensively since the 1950s, nobody seems to have thought of that (i.e., physics of dopants) as a rare fluctuation effect!

In modern parlance, Lifshitz considered the problem with purely off-diagonal (hopping) disorder, whereas Anderson had dealt with purely diagonal (onsite energy) disorder. Many short and intermediate distance properties are better captured by Lifshitz's model of the impurity band, especially at low dopant densities. In his analysis, Lifshitz found that band tails were formed as a result of resonance between pairs of sites that happened to be much closer than average, a rare fluctuation effect. An analogous study of band tails in the Anderson model can also be performed, leading to Lifshitz tails on rare configurations of clusters consisting of resonant sites, with many analytic results being arrived at near the band edge[5-6]. Soon thereafter, band tails were studied in the high-density limit by Halperin and Lax using minimum counting methods[7].

The next major appearance of rare fluctuation effects in condensed matter appears in two studies of many-body spin systems, by McCoy and collaborators[8-9] and by Griffiths[10]. These studies, which appeared around 1968-69, dealt with classical, Ising models of magnetism with disorder. The former was a study of a two-dimensional square Ising ferromagnet with uniform vertical bonds and random bonds in the horizontal direction, which were perfectly correlated in the vertical dimension - it thus appeared as a contrived model in the classical context. However, the results obtained were quite bizarre - it was found that the magnetic susceptibility diverged due to rare configurations of strong bonds in a finite region in the paramagnetic phase above the true thermodynamic phase transition temperature. Griffiths, on the other hand, was studying an Ising ferromagnet in any dimension with randomly diluted bonds. He showed that the existence of rare configuration of clusters with fewer than average missing bonds over finite length scales led to essential singularities in the thermodynamics of the system in the paramagnetic phase, between the transition temperature of the pure, undiluted system, and the actual transition temperature of the diluted system. While these weak singularities in the thermodynamics were experimentally undetectable, it was shown several years later[11] that there was a concomitant slow, non-exponential decay in the long time dynamics of such systems. Such effects were also claimed[12] to affect the long time dynamics of Ising spin glasses (where the bonds are ferromagnetic and antiferromagnetic at random with equal probability); while non-exponential relaxation[13-14] has been a hallmark of spin glasses, whether the long time behavior is dominated by rare configurations remains an open issue at present.

Roughly a decade after Griffiths and McCoy, the problem of randomly positioned donors surfaced again[15], this time in the problem of the magnetic ground state of n-doped semiconductors, which could be modeled as a quantum spin-1/2 Heisenberg antiferromagnet with an extremely high degree of disorder (large randomness). This was addressed using an RG scheme that made use of large disorder, leading to bond distributions that were wide on a logarithmic scale[16]. A one-dimensional analog of the same problem allowed an analytic treatment of the thermodynamics to be obtained in the low temperature limit[17], which was later proved to be asymptotically exact[18]. The RG scheme showed how, in the limit of large disorder, the quantum system's behavior exhibits special characteristics that come about essentially because of enslavement of weak couplings by the strong ones, to a degree that is just not present in classical systems[19]. Such strong effects, which were further amenable to analytic RG treatments, rekindled the interest in rare fluctuation effects in quantum models with disorder.

It was recognized that the McCoy-Wu model was quite natural in the context of the one-dimensional random quantum Ising ferromagnet in a transverse magnetic field[20], the quantum mechanics entering through the non-commuting spin operators (S_Z and S_X) along the Ising coupling (z) and the field (x) directions. The path integral representation of the quantum model led to a classical model in one higher dimension (i.e., two), and perfect correlations between the random bonds in one of the dimensions represented simply the time evolution in the path integral representation. It was shown that approaching the ordered phase at zero temperature at low magnetic fields from the paramagnetic phase at high magnetic fields, the magnetic susceptibility diverged before the (quantum) phase transition, because of the effect of rare configurations of strong bonds. This generated a flurry of activity on transverse field quantum Ising models[21,22], including those with random bonds (e.g. the Ising spin glass in a transverse field). At the same time, it was recognized that similar effects were present in random, dimerized, quantum antiferromagnetic spin chains[23].

The quantum spin glass models in $d = 2$ and higher were not amenable to analytic study, but Monte Carlo simulations[24] showed that the divergences of the non-linear (spin glass) susceptibility in the paramagnetic phase due to rare configurations continued to occur in $d = 2$ and $d = 3$ spatial dimensions. This strong effect in the thermodynamics of the quantum system, as opposed to rather weak effects in the classical system, can be physically understood by considering the path integral mapping of the quantum into a $(d+1)$ dimensional classical system: while the classical random bond model has defects that are point-like in character, the quantum version has line defects (because of the perfect bond-correlation along the time direction).

While these studies showed quite convincingly the overall significant effect of rare configurations for the quantum Ising spin glass, it was after a very detailed numerical effort involving large amounts of computer time. Looking at details as a function of size or geometry of clusters which taken together give rise to the singular behavior was not deemed possible; this situation remains true today.

In the intervening years, there was not much further investigation of Lifshitz-tail[4–6] ideas in the context of the Anderson model[1]; rather the excitement was concentrated mainly on the original issue raised by Anderson, namely, the extended-localized state (or the metal-insulator) transition. Starting with the pioneering numerical efforts of Kramer and MacKinnon[25] and of McMillan and coworkers[26], the study of the metal-insulator transition for non-interacting electrons in a random potential with different symmetry properties (orthogonal, as in potential scattering, unitary as with a magnetic field; or symplectic as with spin-orbit scattering), became a virtual industry, with the most complete work done by Slevin, Ohtsuki and coworkers[27–28]. So it was a complete surprise when our recent investigations[29] pointed out some rather remarkable, singular behavior in Anderson's original model, which had escaped attention for fifty years since it had been proposed. We elaborate upon our findings in the next section.

2. Anderson's 1958 Model Revisited

Anderson[1] considered the one-electron Hamiltonian:

$$H = \sum_i \epsilon_i c_i^\dagger c_i + \sum_{<ij>} V c_i^\dagger c_j, \qquad (1)$$

where i and j denote sites on a simple cubic lattice, ϵ_i are independent random variables drawn from a uniform probability distribution

$$P_U(\epsilon) = 1/W \text{ for } W/2 < \epsilon < W/2, \text{ and zero otherwise}, \qquad (2)$$

and sites i and j within the angular brackets $<ij>$ are nearest neighbors. We will generalize our study slightly, by considering some other probability distributions which are symmetric around zero (see below) and including dimensions $d = 1, 2$ and 3, but restricting our attention to d-dimensional hypercubic lattices. The model is characterized by the dimensionless disorder, W/V.

In three dimensions, for $W = 0$, the density of states (DOS) is the familiar DOS of the nearest neighbor tight-binding model on a simple cubic lattice extending from $E = -6V$ to $+6V$; needless to say, all electronic states are extended, being of the Bloch form. As W/V is increased from zero, states near the band edges get localized, while those near the band center remain extended. Increasing W/V further pushes the mobility edge [at $\pm E_c(W)$] separating the localized tail from the extended center closer to the band center at $E = 0$, until at $W \sim 16.5V$, i.e. for a disorder width of order the original bandwidth $(12V)$, all states become localized [6]. In $d = 1$ and 2, as discussed earlier, all states become localized for any nonzero W/V.

The other distributions we will discuss are Gaussian:

$$P_G(\epsilon) = (2\pi\sigma^2)^{-1/2} exp(-\epsilon^2/2\sigma^2), \qquad (3)$$

where $\sigma = W/\sqrt{12}$ gives the same variance as P_U, the triangular distribution:

$$P_T(\epsilon) = 2(1 - 2|\epsilon|/W)/W \text{ for } |\epsilon| < W/2,), 0 \text{ otherwise}, \qquad (4)$$

as well as a distribution with a power-law behavior at the edges $(\pm W/2)$ and zero density of states at the origin:

$$P_{PL}(\epsilon) \propto |\epsilon|/(W^2 - 4\epsilon^2)^{\alpha-1} \text{ for } |\epsilon| < W/2, 0 \text{ otherwise} \qquad (5)$$

In Eq. 5 $\alpha > 0$, so that the distribution is integrable.

Fig. 1 depicts the first three of these distributions of the on-site energy, given by Eq. 2-4. These distributions will help illustrate our findings and the conditions under which we get them.

In numerical studies of Anderson localization, two main quantities that have proved useful to calculate the localization length at any energy E, $\xi(E)$, have been the inverse participation ratio[29] $\langle IPR(E) \rangle$, and the Lyapunov exponent[6] $Ly(E)$. The former is easily defined for any wavefunction $|\Psi> = \sum_i c_i |i>$, as

$$IPR_\Psi = (\sum_i c_i^4)/(\sum_i c_i^2)^2. \qquad (6)$$

Then

$$\langle IPR(E)\rangle = \langle IPR_\Psi \delta(E - E_\Psi)\rangle, \quad (7)$$

where the average is over states Ψ that are at an energy in a small window around E, which can be made arbitrarily small in the thermodynamic limit, when the typical energy spacing scales as $1/L^d$, where L is the linear dimension of the system. For a state that is localized on a length scale ξ, it is easily seen that $\langle IPR\rangle \sim \xi^{-d}$. For an extended state, on the other hand, $\langle IPR\rangle \sim L^{-d}$, where L is the linear dimension of the system.

The Lyapunov exponent at energy E is computed for systems of finite cross-section (M^{d-1}) and long length in one dimension, which is taken to the thermodynamic limit. It is defined in terms of the eigenvalues of the transfer matrix connecting the amplitudes of a wavefunction at position i and $(i+1)$ with those at $(i+1)$ and $(i+2)$. In dimensions $d > 1$, one has to evaluate it for different M and take the limit as $M \to \infty$. For our purpose, it suffices to note that the limit, $Ly(E)$ characterizes the exponential decay of the wavefunction at long distances, at an energy E, and therefore,

$$Ly(E) \sim 1/\xi(E) \quad (8)$$

On the basis of (7) and (8), one expects therefore that $\langle IPR(E)\rangle \sim [Ly(E)]^d$, and in particular, for $d = 1$, they should be proportional to each other.

In the next section, we describe the results of our numerical investigations, by using $d = 1$, where states are localized at all energies for nonzero disorder ($W/V > 0$). We discuss the situation in higher dimensions briefly in Section 4.

3. Numerical Results (d=1)

Fig. 2 shows $\langle IPR(E)\rangle$ and $Ly(E)$ as a function of E for the Anderson Model (Eq. 1) with a uniform distribution (Eq. 2), in $d = 1$ for $W/V = 6$. Since all states are localized, results converge rapidly as a function of size, and do not require extensive numerical effort. As can be seen, starting from the band center, both $\langle IPR(E)\rangle$ and $Ly(E)$ increase with $|E|$, indicating a decrease in the localization length with increasing $|E|$, as may be expected (band tails localize first even in $d = 3$). However, as one goes to larger $|E|$, $Ly(E)$ continues to rise, but $\langle IPR(E)\rangle$ goes through a maximum, and decreases, apparently going to zero at the band

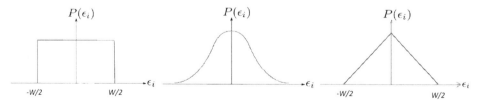

Fig. 1. From left to right, uniform, Gaussian and triangular distribution of disorder.

edge. This is consistent with Lifshitz's considerations —the states at the absolute band edge for finite W [given by $E_B = \pm(W/2 + 2dV)$ for hypercubic lattices in d-dimensions] are due to states on rare clusters of sites with on-site energies all at the edge of the disorder probability distribution, i.e. for $\epsilon_i = \pm W/2$ for all sites. Of course, such a configuration is exponentially rare in the size of the cluster $\sim exp[-cL^d]$ where L is the linear dimension of the cluster, and has a vanishing probability in the thermodynamic limit. Of course, in such a case, the eigenstates are spread all over the cluster since there is no disorder, and would therefore have an $IPR \sim 1/L^d$. Particle in a box considerations suggest that the energy of these states would be $\sim 1/L^2$ away from the true band edge, implying that in the asymptotic region, $\langle IPR(E) \rangle \sim |E_B - E|^{d/2}$, which $\to 0$ as $E \to E_B$. Clearly in this (Lifshitz) regime, the extent of the wavefunction (measured by IPR) and the exponential decay length[measured by $Ly(E)$] are not the same; one has *two* length scales describing the electronic wavefunction.

What is more striking than even the downturn is the manner in which the IPR changes course: there is a clear cusp in the data, implying non-analytic behavior. Furthermore, this non-analytic downturn happens at an energy E_R, where the density of states (DOS) is a maximum, not deep in the exponential tail of the DOS, as shown in Fig. 3. In fact, the DOS also displays a cusp at the same energy. (This does not contradict the result of Edwards and Thouless[31] who showed that the DOS is analytic in an energy interval $(W/2 - 2dV)$ around the band center; the singular behavior we see is outside this region). The cusp-like behavior is seen for intermediate and large (W/V); in $d = 1$, we see it clearly for all $W/V \geq 3.8$. We remark that $Ly(E)$ does not show any sign of the singular behaviour at E_R.

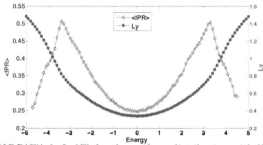

Fig. 2. $\langle IPR(E) \rangle$ & $Ly(E)$ for the uniform distribution with $W/V = 6$.

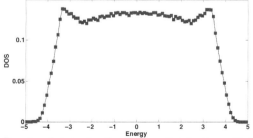

Fig. 3. Density of states for the uniform distribution with $W/V = 6$.

As explained in Ref. 29, this singularity appears to be a consequence of the fact that typical Anderson localized states with a central site do not exist beyond E_R; the only states that are present beyond it are states that involve resonant tunneling between two or more nearest neighbor sites. In the Anderson model, for large disorder W, typical Anderson localized states have their energies perturbed by $O(V^2/W)$, whereas resonant states on two or more sites have energies shifted by $O(V)$; increasing the number of resonant sites only changes the prefactor of the $O(V)$ term, not the functional dependence. As a result, resonant states on sites near the disorder band edge pull out of the quasi-continuum of the typical Anderson-localized states, leaving behind non-analyticity in the density and nature of electronic states at the edge.

E_R thus appears to be the beginning of the Lifshitz tail, whose precise nature depends on short-distance rather than long-distance physics. Furthermore, since the smallest resonant cluster consists of 2 sites, its probability is not exponentially small, but only $O(V/W)$ for large W, so the number of resonant states beyond E_R can be significant. For the case of the uniform distribution in $d = 1$, we find[29] that the maximum percentage of states beyond E_R is $\approx 17\%$ (for $W/V \approx 3.8$). This fraction of tail states can hardly be called rare; hence the quotation marks in the title of this paper. We note in passing that we have shown [30] by explicit analytic calculation for a two-site Anderson model, that such cusp-like singularities exist in both $\langle IPR(E) \rangle$ and $DOS(E)$.

The non-analytic behavior characterizing the separation of resonant states from typical Anderson localized states appears to exist for all forms of bounded disorder. Fig. 4 shows $\langle IPR \rangle$ and DOS as a function of E for the triangular distribution (Eq. 4) for $W/V = 6$. Again the sharp maximum in $\langle IPR \rangle$ occurs at a DOS that is not exponentially small, despite the smoother edge to the probability distribution of the disorder. On the other hand, for the Gaussian distribution of disorder (Eq. 3), we find that both $Ly(E)$ and $\langle IPR(E) \rangle$ increase monotonically with $|E|$, with no sign of a turnaround or strong singularity for up to values of $|E|$ where the DOS is down to 10^{-4} of its value at the band center. This is consistent with Wegner's result[33] that the DOS is analytic at all E for Gaussian disorder.

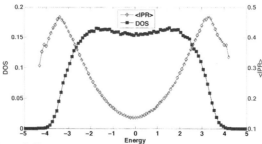

Fig. 4. DOS and $\langle IPR(E) \rangle$ for the triangular distribution of onsite energies with $W/V = 6$.

For distributions like that in Eq. 5 with several distinct edges ($|E| = W/2$, and $E = 0$), we find that the system develops more singular points dividing typical and resonant eigenstates. This is currently under further investigation.

On the basis of the above findings, we conclude that the Anderson model with bounded disorder in $d = 1$, when all states are localized, has a critical energy E_R at least at large disorder, which demarcates the transition from typical Anderson localized states with a central site, to states that necessarily involve resonance between two or more nearest neighbor sites. This is depicted in Fig. 5. When these resonant Lifshitz-like states occur at an energy that is forbidden to typical Anderson localized states, they maintain their strongly resonant character. On the other hand, when they occur at energies in the range allowed to typical Anderson localized states, they mix with the more abundant typical states, and lose their special resonant character. In that sense, this phenomenon appears to be analogous to the fact that localized states can only exist at energies where extended states do not; where they do, they mix and lose their localized character. Whether there is a further hierarchy among resonant states on different number of sites, and whether these are accompanied by (presumably weaker) singularities or not, remains a topic for further investigation.

4. Higher Dimensions and Concluding Remarks

In conclusion, we have shown that Anderson's original model of localization exhibits behavior that is richer than believed, so far. for the uniform distribution, as well as other bounded distributions of diagonal disorder, in the localized phase, the inverse participating ratio characterizing the extent of localized states exhibits a distinct, non-analytic behavior, which is to some extent reflected in the density of states as well. The energy E_R at which this non-analyticity occurs separates typical Anderson localized states from states that have a resonant character on various length scales.

This phenomenon, described above in Section 3 for $d = 1$ is not special to one dimension. As shown in Ref. 29, cusp-like singularities exist for large disorder in $d = 2$, and also in $d = 3$, in the insulating phase. Of course, detailed numbers are dimension dependent. It is interesting to speculate whether in $d = 3$, the existence of a second critical energy E_C denoting the transition from extended to localized behavior will lead to an interplay of some sort. For example, would $|E_R|$ always be bigger than $|E_C|$ for all W, or could there be a transition from extended states

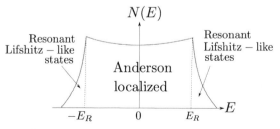

Fig. 5. A schematic illustration showing the occurence of different kinds of wavefunctions in the insulating phase.

directly into localized, resonant states as a function of energy for some value of the disorder? If so, would that be any different from the standard localization transition? In this regard, it is interesting to note that numerical investigations[6,25,34] of the localization transition in the Anderson model in $d = 3$ have found universal behavior for the transition at the band center ($E = 0$) as a function of disorder (W), but have found non-universal results with similar size systems, when they study the transition at fixed W as a function of energy (E). It is tempting to speculate that this may be connected to the change in the nature of the localized state as a function of energy at E_R, which is quite distinct from the true band edge E_B. Clearly, further research needs to be done to determine if the speculation has merit.

Another interesting result of our findings is that the original Anderson model[1] provides a platform for an in-depth study of rare fluctuation physics. While analytic methods are most useful in the asymptotic exponential tail, where numerical methods are of little use, numerical methods can help provide detailed results due to resonant clusters of small finite sizes. If analytic methods can help decipher the nature of further (weaker) singularities due to the transitions from n-site clusters to $(n+1)$-site clusters that may exist as one descends down the tail of the distribution, there is the possibility that numerical methods may be able to pick them up quantitatively. This is because Anderson localization, being a single particle problem, requires time that grows as a power of the system size. Thus it is amenable to a much more intensive numerical investigation than many-body Hamiltonians with exponential growth of the Hilbert space with size. In addition, since most of the physics lies in the insulating phase, the sizes that will need to be considered will likely also be manageable.

Acknowledgments

We thank Dr. Zlatko Papic for help with figures 1 and 5.

References

1. P. W. Anderson, Physical Review **109**, 1492 (1958).
2. N. F. Mott Jour. Non-Crystalline Solids, **1**, 1 (1968).
3. N. F. Mott and W. D. Twose, Adv. Phys. **10**, 107 (1961).
4. E. Abrahams, P. W. Anderson, D. C. Licciardello and T. V. Ramakrishnan, Physical Review Letters **42**, 673 (1979).
5. I. M. Lifshitz, Adv. Phys. **13**, 483 (1964); Soviet Physics Uspekhi **7**, 549 (1965).
6. B. Kramer and A. MacKinnon, Rep. Prog. Phys. **56**, 1469 (1993).
7. B. I. Halperin and Melvin Lax, Physical Review **148**, 722 (1966); ibid. **153**, 802 (1967).
8. B. M. McCoy and T. T. Wu, Physical Review **176**, B631 (1968); ibid. **188**, 982 (1969).
9. B. M. McCoy, Physical Review Letters **23**, 383 (1969); Physical Review **188**, 1014 (1969).
10. R. B. Griffiths, Physical Review Letters **23**, 17 (1969).
11. D. Dhar and M. Barma, Jour. Stat. Phys. **22**, 259 (1980); D. Dhar in Stochastic Processes: Formalism and Applications, G. S. Aggarwal and S. Dattagupta, editors (Springer, Berlin, 1983).

12. M. Randeria, J. P. Sethna, and R. G. Palmer, Physical Review Letters **54**, 1321 (1985).
13. A. Ogielski, Physical Review B **32**, 7384 (1985).
14. P.Svedlindh, P.Granberg, P. Nordblad, L. Lundgren and H. S. Chen, Physical Review B **35**, 268 (1987); K. Gunnarsson, P. Svedlindh, P. Nordblad, L. Lundgren, H. Hruga and A. Ito, Physical Review Letters **61**, 754 (1988).
15. R. N. Bhatt and T. M. Rice, Philos. Mag. **42B**, 859 (1980); M. Rosso, Physical Review Letters **44**, 1541 (1980); R. N. Bhatt, Physical Review Letters 48, 707 (1982).
16. R. N. Bhatt and P. A. Lee, Bull. Am. Phys. Soc., **25**, 206 (1980); Journal of Applied Physics **52**, 1703 (1981); Physical Review Letters **48**, 344 (1982).
17. S. K. Ma, C. Dasgupta and C. K. Hu, Physical Review Letters **43**, 1899 (1979); C. Dasgupta and S. K. Ma, Physical Review B **22**, 1305 (1980)
18. D S Fisher, Physical Review B **50**, 3799 (1994).
19. R.N. Bhatt, Physica **109-110** B&C, 2145 (1982)
20. R. Shankar and G. Murthy, Physical Review B **36**, 536 (1987); D. S. Fisher, Physical Review Letters **69**, 534 (1992); Physical Review B **51**, 6411 (1995).
21. For a review, see Quantum Ising Phases and Transitions in Transverse Ising Models, B. K. Chakrabarti, A. Dutta and P.Sen (Springer, Berlin, 1996).
22. R. N. Bhatt, in Spin Glasses and Random Fields, ed. A. P. Young (World Scientific, 1997).
23. R.A. Hyman, K. Yang, R.N. Bhatt and S.M. Girvin, Physical Review Letters **76**, 839 (1996)
24. H. Rieger and A. P. Young, Physical Review B **54**, 3328 (1996); Muyu Guo, R. N. Bhatt and David A. Huse, Physical Review B **54**, 3336 (1996).
25. A. MacKinnon and B. Kramer, Physical Review Letters **47**, 1546 (1981).
26. W. L. McMillan, Physical Review B **31**, 344 (1985); Avinash Singh and W. L. McMillan, Jour. of Physics C **17**, 2097 (1985).
27. K. Slevin and T. Ohtsuki, Physical Review Letters **78**, 4083 (1997).
28. Y. Asada, K. Slevin, and T. Ohtsuki. Physical Review Letters **89**, 256601 (2002).
29. S. Johri and R. N. Bhatt, arXiv1106.1131 (submitted for publication).
30. F. Evers and A. D. Mirlin, Reviews of Modern Physics **80**, 1355 (2008).
31. J. T. Edwards and D. J. Thouless, Journal of Physics C **4**, 453 (1971).
32. S. Johri and R. N. Bhatt (in preparation)
33. F. Wegner, Z. Phys. B **44**, 9 (1981).
34. B. Kramer, K. Broderix, A. MacKinnon and M. Schreiber, Physica A **167**, 163 (1990).

EFFECT OF ELECTRON-ELECTRON INTERACTION NEAR THE METAL-INSULATOR TRANSITION IN DOPED SEMICONDUCTORS STUDIED WITHIN THE LOCAL DENSITY APPROXIMATION

YOSUKE HARASHIMA and KEITH SLEVIN*

*Department of Physics, Graduate School of Science, Osaka University,
1-1 Machikaneyama, Toyonaka, Osaka 560-0043, Japan
slevin@phys.sci.osaka-u.ac.jp

Revised 4 April 2012

We report a numerical analysis of Anderson localization in a model of a doped semiconductor. The model incorporates the disorder arising from the random spatial distribution of the donor impurities and takes account of the electron-electron interactions between the carriers using density functional theory in the local density approximation. Preliminary results suggest that the model exhibits a metal-insulator transition.

Keywords: disordered system; metal-insulator transition; density functional theory; Anderson localization.

PACS numbers: 71.23.An, 71.30.+h, 71.15.Mb

1. Introduction

In semiconductors a zero temperature metal-insulator transition is observed as a function of doping concentration. For samples with concentrations below a critical concentration, the conductivity extrapolated to zero temperature is found to be zero. For samples with concentrations exceeding this critical concentration, the zero temperature limit of the conductivity is finite.[1] One well studied example is phosphor doped silicon (Si:P) (see Ref. 2 for a recent review). The relative importance of the roles that electron-electron interactions and disorder play in this transition is still not clear. The Coulomb interaction between the electrons leads us to expect that the impurity band is split into upper and lower Hubbard bands and that the transition is associated with a closing of the Hubbard gap. However, this ignores the effect of the disorder that arises from the random spatial distribution of the donor impurities and the possibility of Anderson localization.[3] This paper is a preliminary report of numerical simulations designed to address this issue.

2. Model

As a simple model of a doped semiconductor we consider an effective medium with electron effective mass m_e^* and dielectric constant ε_r equal to those of the host semiconductor crystal. In this effective medium N donor impurities are randomly distributed in space. Since we have mind phosphor in silicon, we assume that each donor supplies one electron and as a result has a net charge of $+e$. This is the only property of the donor which enters our model. There are an equal number of electrons so that the total charge is zero. The electrons interact with the donors through the Coulomb interaction. The random spatial distribution of the donors thus produces a random potential in which the electrons move. At the same time the electrons interact with each other via the Coulomb interaction. The Hamiltonian of this system is

$$\mathcal{H} = -\frac{1}{2m_e^*} \sum_{i=1}^{N} \nabla_i^2 - \frac{1}{\varepsilon_r} \sum_{i,I=1}^{N} \frac{1}{|\vec{r}_i - \vec{R}_I|} + \frac{1}{2\varepsilon_r} \sum_{i \neq j} \frac{1}{|\vec{r}_i - \vec{r}_j|} \,. \tag{1}$$

Here, Hartree atomic units are used. The positions of the donor impurities are denoted by $\{\vec{R}_I\}$. The first term is the kinetic energy of the electrons, the second term describes the interaction of the electrons with the donor impurities, and the third term describes the interaction between the electrons. A fourth term describing the mutual Coulomb interaction between the donor impurities should also be included if the correct total energy of the system is required. However, since the positions of the donor impurities, while random, are fixed, this contribution to the energy does not play any role in the following discussion and is, therefore, omitted.

To deal with the electron-electron interaction we use density functional theory[4] and solve the Kohn-Sham equations[5] that describe an auxiliary one-electron problem that has the same ground state density as the interacting problem of Eq. (1). The Kohn-Sham equations are

$$\left(-\frac{1}{2m_e^*} \nabla^2 + V_{\text{eff}}\left[n\left(\vec{r}\right)\right] \right) \phi_i\left(\vec{r}\right) = \epsilon_i \phi_i\left(\vec{r}\right) \quad (i = 1, \ldots, N) \,, \tag{2}$$

where

$$V_{\text{eff}}\left[n\left(\vec{r}\right)\right] = -\frac{1}{\varepsilon_r} \sum_{I=1}^{N} \frac{1}{|\vec{r} - \vec{R}_I|} + \frac{1}{\varepsilon_r} \int d^3 r' \frac{n\left(\vec{r}'\right)}{|\vec{r} - \vec{r}'|} + V_{\text{XC}}\left[n\left(\vec{r}\right)\right] \,. \tag{3}$$

The number density of the electrons is

$$n\left(\vec{r}\right) = \sum_{i=1}^{N} |\phi_i\left(\vec{r}\right)|^2 \,. \tag{4}$$

Periodic boundary conditions are imposed. In this model the dependence on the medium enters only through the effective mass and dielectric constant. Having in mind silicon as the host semiconductor we set

$$m_e^* = 0.32 \,, \quad \varepsilon_r = 12.0 \,. \tag{5}$$

The exchange-correlation potential appearing in the Kohn-Sham equations is given by the functional derivative of the exchange-correlation energy with respect to the number density of electrons

$$V_{\text{XC}}(\vec{r}) = \frac{\delta E_{\text{XC}}}{\delta n(\vec{r})} . \qquad (6)$$

While in principle the Kohn-Sham equations are exact, in practice the exact form of the exchange-correlation potential is not known and an approximation is required. In this work, we use the local density approximation (LDA) in which the functional is approximated as

$$E_{\text{XC}} \approx E_{\text{XC}}^{\text{LDA}} \equiv \int d^3 r \epsilon_{\text{XC}}(n(\vec{r})) n(\vec{r}) . \qquad (7)$$

In this preliminary work, we assume complete spin polarization.[a] We use the form of ϵ_{XC} given in Eq. (2) of Ref. 6 (with spin polarization $\zeta = 1$) though with the parameter values given in Ref. 7 rather than Ref. 6.

In the literature expressions for the exchange-correlation potential are given for electrons in free space whereas we are considering an effective medium. To map the expressions in the literature to the formulae we require here, we re-scale lengths and energies according to the formulae

$$\tilde{\vec{r}} = (m_e^*/\varepsilon_r) \vec{r} , \quad \tilde{E} = (\varepsilon_r^2/m_e^*) E . \qquad (8)$$

After this re-scaling we have

$$V_{\text{XC}}[n(\vec{r})] = \frac{m_e^*}{\varepsilon_r^2} \cdot \tilde{V}_{\text{XC}}\left[\tilde{n}(\tilde{\vec{r}})\right] , \qquad (9)$$

where

$$\tilde{n}(\tilde{\vec{r}}) = (\varepsilon_r/m_e^*)^3 n(\vec{r}) , \qquad (10)$$

and $\tilde{V}_{\text{XC}}$ is the exchange-correlation potential found in the literature.

In Hartree atomic units the unit of length is the Bohr $a_0 = \varepsilon_0 h^2/(\pi m_e e^2) \approx 5.292 \times 10^{-11}$m. We use this as the unit of length throughout. The simulations are performed by generating an ensemble of cubic samples of linear dimension L. In each sample, donor impurities are randomly distributed on a cubic lattice with lattice constant 36 Bohr. This prevents impurities being positioned un-physically close together by chance. (We do not allow two impurities to be at the same site on this lattice.) The volume of the system is $V = L^3$ and the donor concentration $n_D = N/V$.

For numerical purposes the continuous description above is replaced by a discrete description on a real-space grid with spacing a. Derivatives are replaced by next

---

[a] A calculation that assumes either zero or complete spin polarization is numerically less demanding than a calculation using the local spin density approximation but excludes the possibility of a Hubbard gap from consideration. We do not assume zero polarization because double occupation of the impurities is unlikely in the localized phase.

nearest neighbor finite difference approximations. The resulting matrices and vectors have dimension equal to the number of grid points $(L/a)^3$. When calculating the results shown below we used a grid spacing of

$$a = 18 \text{ Bohr}. \qquad (11)$$

For comparison the effective Bohr radius for an electron in the conduction band of silicon is $a_0^* = \varepsilon_r/m_e$ Bohr ≈ 37.5 Bohr. The potential due to the positive donor impurities is calculated by expressing the charge density of the impurities as a Fourier series. A cut-off is imposed on the wavenumbers so that the number of terms in this series is equal to the number of points of the real-space grid. In effect, this replaces the delta-functions of the charge density of the impurities with an approximate smooth charge density. Poisson's equation is solved exactly for this approximate density and the corresponding potential obtained using an inverse Fourier transform. This calculation need only be performed once for a given impurity configuration. The Hartree like term in Eq. (3) is evaluated in a similar way. This latter calculation needs to be repeated for each iteration. Within the LDA the real space finite difference approximation of the Kohn-Sham equations yields a Hamiltonian matrix that is sparse. The N Kohn-Sham orbitals of lowest energy are found using the JADAMILU sparse matrix library.[8] The self consistent solution of the Kohn-Sham equations is found by iteration.

3. Result

To determine the nature, localized or extended, of the Kohn-Sham wavefunctions we use multi-fractal finite size scaling[9,10] (MFSS) of the wavefunction intensity distribution. The system is divided into boxes of linear dimension l and the Kohn-Sham wavefunction intensity is coarse grained

$$\mu_k \equiv \int_{k\text{th box}} d^3r \, |\phi_i(\vec{r})|^2 . \qquad (12)$$

From these coarse grained intensities a random variable α is defined

$$\alpha \equiv \frac{\ln \mu}{\ln \lambda}, \qquad (13)$$

where λ is the ratio of box and system sizes

$$\lambda \equiv \frac{l}{L}. \qquad (14)$$

As described in Refs. 9 and 10, for Anderson's model of localization[3] the distribution of α is scale invariant at the Anderson transition provided λ is held fixed. The distribution shifts to smaller (larger) values in the metallic (localized) phases as the systems size increases. We expect similar considerations to apply to the intensities of the Kohn-Sham wavefunctions. In what follows we focus on the generalized multi-fractal exponent $\tilde{\alpha}_0$ for the highest occupied Kohn-Sham orbital. We refer the reader to Eqs. (6), (7) and (19) of Ref. 10 for the definition of this quantity. The ensemble

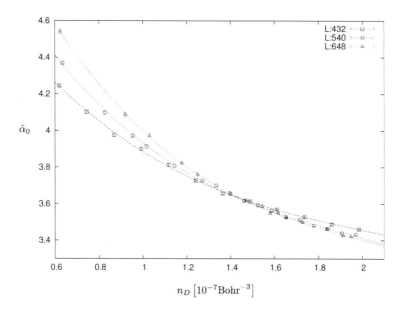

Fig. 1. The value of generalized multi-fractal exponent $\tilde{\alpha}_0$ as a function of donor impurity concentration for $L = 432$, 540, and 648 Bohr. The number of samples simulated varies between 494 and 1004.

average in the definition given there is estimated using an average over samples in the usual way. The precision of this estimate is determined using the formulae given in Table II of Ref. 10.

In Fig.1 we plot the donor concentration dependence of $\tilde{\alpha}_0$ for three different system sizes. Least squares fits of third order polynomials have been made to the data for each system size. For low concentration behavior typical of Anderson localized wavefunctions is seen, i.e. we see shifts to larger values as the system size increases. For high concentrations the system size dependence is less pronounced but there seems to be a shift to smaller values with increasing system size. (Note there is a lower bound of $\tilde{\alpha}_0 \geq 3$ set by wavefunction normalization.) The polynomial fits show a common crossing point around $1.4 \times 10^{-7} \text{Bohr}^{-3}$. The data, though certainly very preliminary, suggest a metal-insulator transition near this value. For comparison the experimental critical concentration for the metal-insulator transition in Si:P is $\simeq 5.2 \times 10^{-7} \text{Bohr}^{-3}$.[2] It is possible that this discrepancy arises from our simplifying assumption of complete spin polarization.

Examination of the electron density (not shown here) for the highest occupied Kohn-Sham wavefunction of typical samples reveals the following. For low concentrations $\sim 0.6 \times 10^{-7} \text{Bohr}^{-3}$ the Kohn-Sham wavefunctions resemble molecular orbitals on clusters of two or three impurities. As the concentration is increased towards $\sim 1.4 \times 10^{-7} \text{Bohr}^{-3}$ the Kohn-Sham wavefunctions spread out over more impurities but remain localized. For higher concentration $\sim 1.8 \times 10^{-7} \text{Bohr}^{-3}$ the Kohn-Sham wavefunctions are extended across the entire sample.

4. Conclusion

We have presented a model for a doped semiconductor that includes both the random spatial distribution of the donor impurities and the interaction between the carrier electrons, and whose numerical analysis allows the metal-insulator transition to be studied. Preliminary results indicate that the model does exhibit a metal-insulator transition. Since in Eq. (1) the properties of the semi-conductor enter only through the effective mass and relative dielectric constant, we automatically recover the behavior described in Fig. 1 of Edwards and Sienko[11] that to a good approximation the critical concentrations in various semiconductors obey

$$n_c^{1/3} a_0^* = \text{constant} . \quad (15)$$

However, our preliminary estimate of this constant, ≈ 0.19, differs significantly from the observed value of ≈ 0.26. As mentioned already, it is possible that this discrepancy will be resolved when our calculations are extended to properly take account of the electron spin.

In future work we hope to shed light on the relative importance of the roles of disorder and interaction in this transition by making a careful comparison of the critical properties of this model with the known properties of the Anderson transition in Anderson's model of localization for non-interacting electrons. Due to space limitations we also defer a discussion of previous work.[12-14]

Acknowledgments

This work was supported in part by Global COE Program (Core Research and Engineering of Advanced Materials-Interdisciplinary Education Center for Materials Science), MEXT, Japan. We thank Hisazumi Akai for fruitful discussion.

References

1. N. F. Mott, *Metal-Insulator Transitions*, 2nd ed. (Taylor & Francis, London, 1990).
2. H. von Lohneysen, Annalen der Physik **523**, 599 (2011).
3. P. W. Anderson, Physical Review **109**, 1492 (1958).
4. P. Hohenberg and W. Kohn, Physical Review **136**, B864 (1964).
5. W. Kohn and L. J. Sham, Physical Review **140**, A1133 (1965).
6. O. Gunnarsson, B. I. Lundqvist, and J. W. Wilkins, Physical Review B **10**, 1319 (1974).
7. J. F. Janak, V. L. Moruzzi, and A. R. Williams, Physical Review B **12**, 1257 (1975).
8. M. Bollhöfer and Y. Notay, Computer Physics Communications **177**, 951 (2007).
9. A. Rodriguez, L. J. Vasquez, K. Slevin, and R. A. Römer, Physical Review Letters **105**, 046403 (2010).
10. A. Rodriguez, L. J. Vasquez, K. Slevin, and R. A. Römer, Physical Review B **84**, 134209 (2011).
11. P. P. Edwards and M. J. Sienko, Physical Review B **17**, 2575 (1978).
12. J. H. Rose, H. B. Shore and L. M. Sander, Physical Review B **21**, 3037 (1980).
13. R. N. Bhatt and T. M. Rice, Physical Review B **23**, 1920 (1981).
14. E. Nielsen, Ph.D. thesis, Princeton University, 2008.

CAN DIFFUSION MODEL LOCALIZATION IN OPEN MEDIA?

CHU-SHUN TIAN

Institute for Advanced Study, Tsinghua University
Beijing, 100084, China
ct@thp.uni-koeln.de

SAI-KIT CHEUNG

Department of Physics, Hong Kong University of Science and Technology
Hong Kong, China

ZHAO-QING ZHANG

Department of Physics and William Mong Institute of Nano Science and Technology
Hong Kong University of Science and Technology
Hong Kong, China
phzzhang@ust.hk

We employed a first-principles theory – the supersymmetric field theory – formulated for wave transport in very general open media to study static transport of waves in quasi-one-dimensional localized samples. We predicted analytically and confirmed numerically that in these systems, localized waves display an unconventional diffusive phenomenon. Different from the prevailing self-consistent local diffusion model, our theory is capable of capturing all disorder-induced resonant transmissions, which give rise to significant enhancement of local diffusion inside a localized sample. Our theory should be able to be generalized to two- and three-dimensional open media, and open a new direction in the study of Anderson localization in open media.

Keywords: Anderson localization; local diffusion; open media.

PACS numbers: 42.25.Dd, 71.23.An

1. Introduction

Anderson localization of waves in open media has been an important but difficult area of study in the past few decades. In realistic experiments, one always needs to allow wave energy to leak out of a disordered sample in order to facilitate measurements. In addition, the interplay between strong localization of waves inside the sample and the open boundaries gives rise to resonant transmissions, which was first pointed out by Azbel about three decades ago.[1] Although these resonant transmissions are rare events, they contribute significantly to the transport of wave energy in the localized sample due to their large transmissions and have found many applications such as the random lasing[2] and optical bistability.[3]

How to theoretically study the interplay between strong localization and open boundaries has been a difficult task. About a decade ago, van Tiggelen et. al[4] proposed a self-consistent local diffusion (SCLD) model to study the transport of localized waves in open media (for a review, see Ref. 5). In this model they generalized the self-consistent localization theory for an infinite medium[5] to a finite-sized medium by imposing a position-dependent diffusion coefficient to take into account the fact that the amount of reduction in diffusion coefficient due to weak localization effects is less for a point near a boundary than a point away from the boundaries. Despite its phenomenological nature, the SCLD model has been applied to various systems during the past decade. However, it was found recently that the dynamic version of SCLD model failed to describe the pulsed microwave measurements of wave transport in quasi-one-dimensional localized samples at long times, when wave energy is mainly stored in long-lived modes.[6] Since such long-lived modes are responsible for the resonant transmissions[1], the failure of the SCLD model at long times indicates its inability to capture resonant transmission states. This suspicion is also supported by the fact that the SCLD model is a local model, whereas the disorder-induced resonant transmissions are highly non-local objects.

In a series of works,[7,8] we formulated a first-principles theory – the supersymmetric field theory – for wave transport in very general open media. We employed it to study static transport of waves in quasi-one-dimensional localized samples. We predicted analytically an unconventional diffusive phenomenon for localized waves in these systems. The analytic prediction was confirmed by numerical simulations. In addition, we showed that the SCLD model is incapable of capturing disorder-induced resonant transmissions. Our findings should be able to be generalized to two- and three-dimensional open media, and open a new direction in the study of Anderson localization in open media.

2. Supersymmetric field theory

Consider the propagation of a scalar wave of frequency ω in a finite-sized random medium embedded in the air background. Both the shape of the interface, C, and the local transmission coefficient on C may be arbitrary. The size of the random medium is much larger than the transport mean free path ℓ. The propagation of waves inside the medium is described by the retarded/advanced Green function, $G^{R,A}_{\omega^2}(\mathbf{r},\mathbf{r}')$, obeying

$$\left\{\nabla^2 + \left(\frac{\omega^\pm}{c}\right)^2 [1+\epsilon(\mathbf{r})]\right\} G^{R,A}_{\omega^2}(\mathbf{r},\mathbf{r}') = \delta(\mathbf{r}-\mathbf{r}'), \qquad (1)$$

where $\omega^\pm = \omega \pm i0^+$ and c is the velocity of waves in the air. The fluctuating dielectric field $\epsilon(\mathbf{r})$ has zero mean and follows a Gaussian δ-correlated law. To make progress, we calculated the propagator, $Y(\mathbf{r},\mathbf{r}';\Omega)$, defined as

$$Y(\mathbf{r},\mathbf{r}';\Omega) = \frac{\omega^2}{2\pi\nu} \overline{G^R_{(\omega+\Omega/2)^2}(\mathbf{r},\mathbf{r}') G^A_{(\omega-\Omega/2)^2}(\mathbf{r}',\mathbf{r})} \qquad (2)$$

with $\Omega \ll \omega$. Here, ν is the density of states per unit volume, and the overline stands for the disorder average. This quantity probes the average wave intensity at $\mathbf{r}$ in the presence of an external source at $\mathbf{r}'$.

The propagator (2) was calculated[7,8] by using the rigorous supersymmetric field theoretic approach.[9] Specifically, $Y(\mathbf{r}, \mathbf{r}'; \Omega)$ was expressed in terms of the functional integral over the supermatrix field, $Q(\mathbf{r})$, i.e.,

$$Y(\mathbf{r}, \mathbf{r}'; \Omega) = \frac{\pi \nu}{2^7} \int D[Q] \text{str}[\sigma_3^{\text{BF}}(1 + \sigma_3^{\text{AR}})(1 - \sigma_3^{\text{TR}}) \\ \times Q(\mathbf{r})(1 - \sigma_3^{\text{AR}})(1 - \sigma_3^{\text{TR}})\sigma_3^{\text{BF}} Q(\mathbf{r}')] e^{-F[Q]}. \quad (3)$$

The Q field represents the low energy modes namely the diffuson and the cooperon (in the presence of time-reversal symmetry) and their interactions. Q is an 8×8 matrix defined on three sectors, each of which accommodates two degrees of freedom. The 'AR'-sector accounts for the analytic structures of advanced/retarded Green function; the 'BF'-sector for the 'supersymmetry', with the diagonal (off-diagonal) entries in this sector being complex (Grassmann) – bosonic (ferminionic) – numbers; the 'TR'-sector for the symmetric properties under time-reversal operations, with Q being diagonal in this sector for the broken time-reversal symmetry. Q is parametrized by $Q = T\sigma_3^{\text{AR}} T^{-1}$, where T takes the matrix value in some coset space corresponding to the orthogonal (time-reversal) and unitary (broken time-reversal) symmetry, respectively. $\sigma_3^X = \text{diag}(1, -1)^X$ is defined on $X(=$'AR', 'BF', 'TR')-sector and 'str' the supertrace.[9] The action,

$$F[Q] = \frac{\pi \nu}{8} \int d\mathbf{r} \,\text{str}\, [D_0(\nabla Q)^2 + 2i\Omega^+ \sigma_3^{\text{AR}} Q], \quad (4)$$

describes Q-field fluctuations, where $D_0 = c\ell/3$ is the Boltzmann diffusion constant. Eq. (3) differs in essence from that of infinite media in the boundary condition satisfied by Q, which is

$$\left(2\zeta Q(\mathbf{r})\nabla_{\mathbf{n}(\mathbf{r})} Q(\mathbf{r}) + [Q(\mathbf{r}), \sigma_3^{\text{AR}}]\right)_\perp \Big|_{\mathbf{r} \in C} = 0. \quad (5)$$

Here, ζ is the extrapolation length determined by the internal reflection coefficient. $\nabla_{\mathbf{n}(\mathbf{r})}$ is the (local) normal derivative on the boundary and the subscript '$\perp$' stands for the off-diagonal entries in the AR-sector. Eqs. (4) and (5) constitute the microscopic theory for transport of localized waves in general finite-sized media.

For fully transparent boundaries, the boundary condition is simplified to

$$Q(\mathbf{r})\big|_{\mathbf{r} \in C} = \sigma_3^{\text{AR}}. \quad (6)$$

Eqs. (3), (4) and (6) are the starting point of studies of wave transport in open media.

3. Local diffusion of waves in open media

Local diffusion equation. — Calculating Eq. (3) explicitly, we found that the propagator $Y(\mathbf{r}, \mathbf{r}'; \Omega)$ solves the following 'local diffusion equation',

$$\{-\nabla D(\mathbf{r}; \Omega) \nabla - i\Omega\} Y(\mathbf{r}, \mathbf{r}'; \Omega) = \delta(\mathbf{r} - \mathbf{r}'). \tag{7}$$

It suggests an unconventional diffusive phenomenon in open media. The key difference from the normal diffusion is that the diffusion coefficient, $D(\mathbf{r}; \Omega)$, is position-dependent: near the boundary, the diffusion coefficient is largely unaffected, i.e., $D(\mathbf{r}; \Omega) \simeq D_0$, but deep inside the medium, D_0 is strongly suppressed down to a value which may be smaller by many orders (see below for example). Physically, a local (position-dependent) diffusion coefficient finds its origin in that in open media, the returning probability is inhomogeneous in space because of wave energy leakage through the boundaries, and as localization corrections to D_0 functionally depends on this probability, the diffusion coefficient becomes inhomogeneous also.

Static local diffusion in quasi-one-dimensional localized samples. — In this case, $\Omega = 0$, the local diffusion equation is reduced to $-\partial_x D(x) \partial_x Y(x, x') = \delta(x - x')$ and the local diffusion coefficient, $D(x)$, was calculated explicitly.[8] Firstly, a novel scaling behavior exhibited by the local diffusion coefficient was found. I.e.,

$$D(x)/D_0 = D_\infty(\lambda), \qquad \lambda = \frac{(L-x)x}{L\xi}, \tag{8}$$

where the scaling function $D_\infty(\lambda) \sim e^{-\lambda}$ for $\lambda \gg 1$. Notice that the sample length, L, is much larger than the localization length, ξ. Secondly, deep inside the sample, i.e., $\xi \ll x$ or $(L - x) \leq L/2$, the local diffusion coefficient was found to increase drastically from the exponential decay,

$$D(x)/D_0 \propto e^{x^2/(L\xi)} e^{-x/\xi}. \tag{9}$$

As presented in Fig. 1, these results were entirely confirmed by numerical simulations of the spatially resolved wave intensity across a randomly layered medium (with the

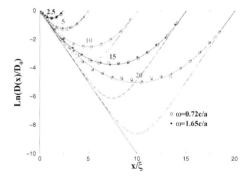

Fig. 1. We computed $D(x)/D_0$ numerically for two wave frequencies, $\omega = 1.65c/a$ (square) and $\omega = 0.72c/a$ (circle), and for five different sample lengths, $L/\xi = 2.5, 5, 10, 15$ and 20. The results were compared with the analytic prediction namely Eq. (9) (solid lines) and that obtained from the SCLD model (dashed lines). (from Ref. 8)

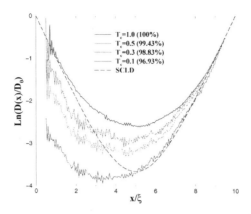

Fig. 2. For the original ensemble ($\omega = 1.65c/a$ and $L/\xi = 10$), we eliminated the high-transmission ($T > T_c$) states and re-computed $D(x)$ (solid lines). The source is placed at $x = 0$. (from Ref. 8)

layer thickness a). Fig. 1 also shows unambiguously that the SCLD model[4] fails in these samples.

Here we would like to make three more remarks. (i) Solving the static local diffusion equation with $D(x)$ given by Eq. (9), we found the average transmission as $\langle T(L) \rangle \propto e^{-L/(4\xi)}$ for $L \gg \xi$. On the other hand, for $T(L) = e^{-\gamma L}$ and γ follows the Gaussian distribution, the typical transmission gives $d\langle \ln T(L)\rangle/dL = -\xi^{-1}$. We see that the localization lengths obtained by the arithmetic mean is four times larger than that obtained by the geometric mean. This is in agreement with the result obtained from the random matrix theory,[10] and is due to that resonant transmissions dominate the average transmission. (ii) In localized samples, the concept of local diffusion is closely associated with the resonant transmissions. In Fig. 2, we showed that the local diffusion breaks down even a tiny fraction of resonant transmissions were eliminated in the numerical calculation. Thus, these rare high-transmission states play important roles in establishing local diffusion and scaling behavior. (iii) Eqs. (8) and (9), as well as the remarks above namely (i) and (ii), remain valid when the time-reversal symmetry is broken.

4. Concluding Remarks

We formulated a first-principles theory – the supersymmetric field theory – for localization in open media and proved the existence of local diffusion in these systems. For static transport of waves in quasi-one-dimensional localized samples, we calculated the (static) local diffusion coefficient explicitly. Our analytic result shows a novel scaling behavior as a result of resonant transmissions, whose roles in the transport of localized waves are decisive. Our result also shows that only if resonant transmissions can be neglected, the prevailing SCLD model is valid.

Theoretical investigations of a number of important problems, which are undergoing intensified experimental studies, may be proceeded along the same line:

- We expect that the dynamic local diffusion coefficient may also display some novel scaling behavior. This may help to understand (i) roles of disordered-induced resonant transmission and local diffusion in the pulsed microwave experiment,[6] (ii) the relation between local diffusion and transport through modes,[11] and (iii) the interplay between local diffusion and weak dissipation.[12]
- We expect that in higher dimensions, criticality and open boundaries may strongly interplay with each other, leading to richer local diffusive phenomena.

These issues are currently under investigations.

Acknowledgments

We thank A. Z. Genack and A. Kamenev for discussions, and A. Altland, D. Basko, A. A. Chabanov, A. De Martino, M. Garst, and T. Nattermann for conversations. Work supported by SFB TR12 of the DFG, by the HK RGC No. 604506, by the NSFC (No. 11174174), and by the Tsinghua University Initiative Scientific Research Program (No. 2011Z02151).

References

1. M. Y. Azbel, Phys. Rev. B **28**, 4106 (1983).
2. H. Cao, Y. G. Zhao, S. T. Ho, E. W. Seelig, Q. H. Wang, and R. P. H. Chang, Phys. Rev. Lett. **82**, 2278 (1999).
3. I. V. Shadrivov, B. Yu. Konstantin, Yu. P. Bliokh, V. Freilikher, and Yu. S. Kivshar, Phys. Rev. Lett. **104**, 123902 (2010).
4. B. A. van Tiggelen, A. Lagendijk, and D. S. Wiersma, Phys. Rev. Lett. **84**, 4333 (2000).
5. P. Wölfle and D. Vollhardt, in *Fifty Years of Anderson Localization*, edited by E. Abrahams (World Scientific, Singapore, 2010).
6. Z. Q. Zhang, A. A. Chabanov, S. K. Cheung, C. H. Wong, and A. Z. Genack, Phys. Rev. B **79**, 144203 (2009).
7. C. Tian, Phys. Rev. B **77**, 064205 (2008).
8. C. S. Tian, S. K. Cheung, and Z. Q. Zhang, Phys. Rev. Lett. **105**, 263905 (2010).
9. K. B. Efetov, Adv. Phys. **32**, 53 (1983).
10. C. W. J. Beenakker, Rev. Mod. Phys. **69**, 731 (1997).
11. J. Wang and A. Z. Genack, Nature **471**, 345 (2011).
12. K. Yu. Bliokh, Yu. P. Bliokh, V. Freilikher, A. Z. Genack, B. Hu, and P. Sebbah, Phys. Rev. Lett. **97**, 243904 (2006).

LOCAL PSEUDOGAPS AND FREE MAGNETIC MOMENTS AT THE ANDERSON METAL-INSULATOR TRANSITION: NUMERICAL SIMULATION USING POWER-LAW BAND RANDOM MATRICES

IMRE VARGA

Elméleti Fizika Tanszék, Budapesti Műszaki és Gazdaságtudományi Egyetem,
H-1521 Budapest, Hungary
varga@phy.bme.hu

STEFAN KETTEMANN

Jacobs University, School of Engineering and Science, Campus Ring 1, 28759 Bremen, Germany
s.kettemann@jacobs-university.de

EDUARDO R. MUCCIOLO

Department of Physics, University of Central Florida, Orlando, FL 32816-2385, USA
mucciolo@physics.ucf.edu

Received 3 January 2012
Revised 31 January 2012

At the Anderson metal-insulator transition the eigenstates develop multifractal fluctuations. Therefore their properties are intermediate between being extended and localized. As a result these wave functions are power-law correlated, which causes a substantial suppression of the local density of states at some random positions, resembling random local pseudogaps at the Fermi energy. Consequently the Kondo screening of magnetic moments is suppressed when a magnetic impurity happens to be at such a position. Due to these unscreened magnetic moments the critical exponents and multifractal dimensions at the metal-insulator transition take their smaller, unitary ensemble values for exchange couplings not exceeding a certain critical value $J^* \approx .3D$, where D is the band width. Here we present numerical calculations of the distribution of Kondo temperatures for the critical Power-law Band Random Matrix (PBRM) ensemble, whose properties are similar to that of the Anderson transition with the advantage of using a continuous parameter for tuning the generalized multifractal dimensions of the eigenstates.

Keywords: Kondo effect; Anderson localization; multifractality.

PACS numbers: 72.10.Fk,72.15.-m,75.20.Hr

1. Introduction

The disorder driven Anderson metal-insulator transition (AMIT) continues to be an interesting topic of current condensed matter physics[1,2,3]. The complex structure of the eigenstates right at the transition point has also been discussed and investigated intensively (see Ref. 3 for details). Despite of many numerical investigations over the past decades the experimental relevance of the multifractal nature of these states has been shown clearly only recently in magnetic semiconductors[4], in ultrasound experiments[5] or STM probing of states at the quantum-Hall transition[6].

The interplay of these unusual, multifractal states with other degrees of freedom have been brought to the focus over the past years. For example this is the key feature in a recent study of the enhancement of the critical transition temperature to the superconducting state[7]. The impact of interaction on dephasing effects in the presence of multifractality has been investigated in Ref. 8.

In the present work we provide numerical evidence of the intimate interplay of two famous, low-temperature, universal phenomena, the Kondo effect and the Anderson metal–insulator transition along the results presented in Ref. 9. We study the coupling of a single impurity placed in a host at the Anderson transition. The host is modeled by a random matrix termed as power-law band random matrix (PBRM). The isolated sample of length L is represented by an $L \times L$ real symmetric (time reversal invariant) matrix whose entries are randomly drawn from a normal distribution with zero mean, $\langle H_{ij}^0 \rangle = 0$, and a variance depending on the distance of the matrix element from the diagonal as[3]

$$\langle (H_{ij}^0)^2 \rangle = \frac{1}{2} \frac{\delta_{ij} + 1}{1 + [\sin(\pi|i-j|/L)/(\pi b/L)]^{2\alpha}}, \qquad (1)$$

where b and α are parameters of the model. In order to reduce finite size effects this expression already incorporates periodic boundary conditions, where Eq. (1) is known as the periodic PBRM model. Nevertheless, for sites far away from each other and the border, i.e., if $1 \ll |i-j| \ll N$, the variance decays with a power law

$$\langle (H_{ij}^0)^2 \rangle \sim \left(\frac{b}{|i-j|} \right)^{2\alpha}. \qquad (2)$$

Field-theoretical considerations and detailed numerical investigations verified[3] that the PBRM model undergoes a transition at $\alpha = 1$ from delocalized states for $\alpha < 1$ to localized states for $\alpha > 1$. This transition shows almost all the key features of the AMIT, including multifractality of eigenfunctions and non-trivial spectral statistics at the critical point. However, the PBRM model does not reproduce mobility edges separating localized and extended states that emerge in the case of the conventional Anderson problem. An important point for the PBRM model, however, is the additional parameter b, which provides a continuous variation of the properties of the eigenstates, as well, as the spectral fluctuations by changing this parameter from zero to infinity. Hence setting $\alpha = 1$ we have a critical line as $0 < b < \infty$, where the limit of $b \gg 1$ corresponds to "metallic-like" behavior which is termed as weak–multifractality while the other limit of $b \ll 1$ shows "insulator-like" behavior that is known as strong-multifractality.

As for the magnetic impurity in such a disordered host we take the model described in[11,12,13]. The host is represented by the PBRM model described above with the choice of $\alpha = 1$. The parameter b is chosen over a wide range: $0.05 < b < 50.0$. The magnetic impurity is placed at every site and the effective Hamiltonian of the coupled system reads as

$$H = H^0 + J\vec{S}\vec{s}(0), \quad (3)$$

where H^0 stands for the electronic system in the disordered potential represented by Eq. (1) and the second term describes the $s-d$ contact interaction between the magnetic impurity, $\vec{S}$, coupled antiferromagnetically to the spin–density of the electronic system, $\vec{s}(0)$, at the place of the impurity. Parameter J is the strength of this coupling.

The Kondo temperature, T_K is defined using the perturbative 'one-loop' Nagaoka-Suhl formula[14]

$$\frac{1}{J} = \frac{1}{2N} \sum_{n=1}^{N} \frac{L^d |\psi_n(\mathbf{r})|^2}{|E_n - E_F|} \tanh\left(\frac{E_n - E_F}{2T_K}\right), \quad (4)$$

where $N \sim L^d$ is the number of states in the band of width D, E_F is the Fermi energy, and $|\psi_n(0)|^2$ is the probability density of the nth eigenstate at the position of the impurity with eigenvalue E_n. Unscreened, free magnetic moments may exist at such positions, where Eq. (4) does have a solution setting $T_K = 0$ if $J < J_c$, where

$$\frac{1}{J_c} = \frac{1}{2N} \sum_{n=1}^{N} \frac{L^d |\psi_n(\mathbf{r})|^2}{|E_n - E_F|}, \quad (5)$$

For a clean metallic system with a flat band, $J_c \sim D/\ln N$, vanishing logarithmically with N, leaving no free moments for any finite J. On the other hand, if the energy eigenstates at the Fermi energy become localized with a localization length ξ, finite local gaps of order $\Delta_I = (\rho \xi^d)^{-1}$ appear at the Fermi energy, cutting off the Kondo renormalization[15] (ρ, average density of states). Therefore, there are with certainty free moments whenever $\Delta_I \gg T_K$, or, equivalently, $J \ll J_c^A \sim D/\ln N_I$, where $N_I = D/\Delta_I$ is the number of localized states with a finite wave function amplitude at the magnetic impurity site.

Therefore in search for the existence of free magnetic moments, first we plot the distribution of the maximum of $1/J_c = \rho_{eff}$ which also serves as a lower cutoff for the coupling constant, J_c, for the formation of the Kondo effect[13]. The quantity is the effective, weighted local density of states (LDOS).

$$\rho_{eff} = \frac{1}{2} \sum_{n=1}^{N} \frac{|\psi_i|^2}{|E_n - E_F|}. \quad (6)$$

Apparently as seen in Fig. 1 in our case the minimum of ρ_{eff} depends on system size in contrast to the findings of 13 (cf. Fig. 5 Inset).

In our previous paper[9], we have derived expressions concerning the critical value of the coupling constant, J_c,

$$J_c/D = \sqrt{1 - d_2(b)}, \qquad (7)$$

This is the critical value of the exchange coupling at which $\sqrt{N}$ magnetic moments remain free. The quantity, d_2 appearing in this formula is called the correlation dimension and plays an important role in many different features of the AMIT[3].

In order to check this relation numerically, we calculated $P(1/J_c)$ for several values of b different system sizes. The results for the case of strong multifractality ($b = 0.1$) are shown in Fig. 1. From those data, we extracted the largest value of J_c beyond which the distribution drops to zero with a Gaussian tail. The resulting J_c as function of b is plotted for different system sizes in Fig. 2. For large system sizes, J_c/D converges to unity at $b = 0$ and decays with a power–law for growing b,

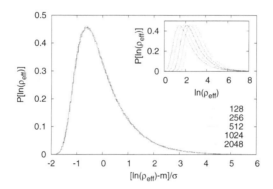

Fig. 1. (Color online) The distribution of $\rho_{eff} \equiv D/J_c$ for $b = 0.1$ for various sizes L. The $\ln \rho_{eff}$ has been rescaled by its mean, m, and variance σ. In the inset the distributions are given as a function of the unscaled variable. Note that the distribution shifts to large values as the system size is increased.

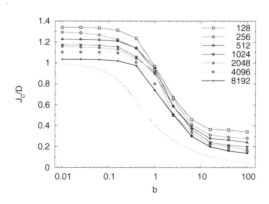

Fig. 2. (Color online) J_c/D, as function of b for various system sizes L. Red solid line: analytical result, Eq. (7).

in good agreement with the analytical result of Eq. (7). As also published in Ref. 9 we found the density of free magnetic moments compared to the total number of magnetic impurities. The results of Ref. 9 have been applied for the case of the critical PRBM, by substituting in

$$P_{\rm FM} = \frac{n_{\rm FM}}{n_{\rm M}} = {\rm Erfc}\left(\sqrt{\frac{\ln\xi}{2(1-d_2(b))}}\frac{J}{D}\right), \qquad (8)$$

$\xi \to L$, which is plotted in Fig. 3 together with the numerical results as function of J/D for various values of parameter b. The probability of finding no solution to Eq. (4) as presented in Fig. 3 seems to be very similar to the one obtained in Ref. 13 (cf. Fig. 4a). In Fig. 3 we show the numerical results for $L = 1024$ and apparently the parameter b produces a similar effect as $1/W$ for the conventional Anderson problem.

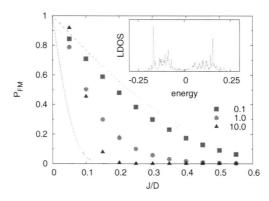

Fig. 3. (Color online) The portion of free magnetic moments $P_{\rm FM} = n_{\rm FM}/n_{\rm M}$, as function of exchange coupling J/D, for various values of b for size $N = 1024$. The analytical result, Eq. (8), $\xi \to L$ and $d = 1$, is plotted for comparison. Inset: a typical plot of the local density of states at a site where the magnetic moment remains unscreened.

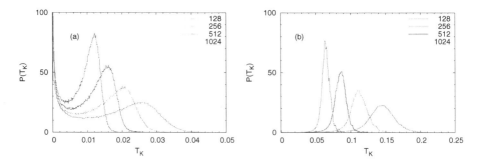

Fig. 4. (Color online) The distribution of the local Kondo temperatures, T_K in arbitrary units, $P(T_K)$ for exchange coupling strengths (a) $J/D = 0.2$, (b) $J/D = 0.9$ for various sizes L using $b = 0.5$.

As it is plotted in the inset of Fig. 3, we can clearly see that for intermediate values of b, the power-law wavefunction correlations yield a finite probability of having a pseudogap with an exponent related to the correlation dimension as d_2.

Finally in Fig. 4 we plot the evolution of the full distribution of the probability density function of the Kondo temperatures, $P(T_K)$, obtained from the solution of Eq. (4) inserting a magnetic impurity at every single site as parameters, b and J change. The bimodal character of $P(T_K)$ is prominent for low value of J/D, while a smearing of the Kondo temperature is detected in the case of strong exchange coupling. The mean and the variance of T_K depend strongly on N.

Acknowledgments

This research was supported by WCU (World Class University) program through the National Research Foundation of Korea funded by the Ministry of Education, Science and Technology(R31-2008-000-10059-0), Division of Advanced Materials Science; it was also supported by the German Research Council under SFB 668, B2, the Alexander von Humboldt Foundation, and the Hungarian Research Fund (OTKA) under K743361 and K75529. ERM and IV thank the WCU AMS and APCTP for their hospitality.

References

1. P. W. Anderson, *Phys. Rev.* **109**, 1492 (1958).
2. D. Belitz and T. R. Kirkpatrick, *Rev. Mod. Phys.* **66**, 261 (1994).
3. F. Evers and A. D. Mirlin, *Rev. Mod. Phys.* **80**, 1355 (2008).
4. A. Richardella, *et al. Science* **327**, 665 (2010).
5. S. Faez, *et al.*, *Phys. Rev. Lett.* **103**, 155703 (2009).
6. K. Hashimoto, *et al.*, *Phys. Rev. Lett.* **101**, 256802 (2008).
7. M. V. Feigel'man, *et al. Phys. Rev. Lett.* **98**, 027001 (2007); M. V. Feigel'man, *et al. Annals of Physics* **325**, 1368 (2010); I.S. Burmistrov, *et al.* arXiv:1102.3323.
8. I.S. Burmistrov, *et al. Ann. Phys.* **326** 1457 (2011).
9. S. Kettemann, E. R. Mucciolo, and I. Varga, *Phys. Rev. Lett.* **103**, 126401 (2009).
10. A. D. Mirlin, F. Evers, I. V. Gornyi, P. M. Ostrovsky, in "50 Years of Anderson Localization", ed. by E. Abrahams (World Scientific, 2010), Chapter 6, pp. 107-150; *Int. J. Mod. Phys.* B **24**, 1577 (2010).
11. S. Kettemann and E. R. Mucciolo, *JETP Lett.* **83**, 240 (2006) [Pis'ma v ZhETF **83**, 284 (2006)]; *Phys. Rev. B* **75**, 184407 (2007).
12. P. S. Cornaglia, D. R. Grempel, and C. A. Balseiro, *Phys. Rev. Lett.* **96**, 117209 (2006).
13. A. Zhuravlev *et al.*, *Phys. Rev. Lett.* **99**, 247202 (2007).
14. Y. Nagaoka, *Phys. Rev.* **138**, 1112 (1965); H. Suhl, *Phys. Rev.* **138**, 515 (1965).
15. S. Kettemann and M. E. Raikh, *Phys. Rev. Lett.* **90**, 146601 (2003).

Localisation 2011
International Journal of Modern Physics: Conference Series
Vol. 11 (2012) 108–113
© World Scientific Publishing Company
DOI: 10.1142/S2010194512005983

FINITE SIZE SCALING OF THE TYPICAL DENSITY OF STATES OF DISORDERED SYSTEMS WITHIN THE KERNEL POLYNOMIAL METHOD

DANIEL JUNG

School of Engineering and Science, Jacobs University Bremen gGmbH,
Campus Ring 1, 28759 Bremen, Germany
d.jung@jacobs-university.de

GERD CZYCHOLL

Institute for Theoretical Physics, University of Bremen,
Otto-Hahn-Allee 1, 28359 Bremen, Germany
czycholl@itp.uni-bremen.de

STEFAN KETTEMANN

School of Engineering and Science, Jacobs University Bremen gGmbH,
Campus Ring 1, 28759 Bremen, Germany
Division of Advanced Materials Science, Pohang University of Science and Technology,
San 31, Hyoja-dong, Nam-gu, Pohang 790-784, South Korea[*]
s.kettemann@jacobs-university.de

We study the (Anderson) metal-insulator transition (MIT) in tight binding models (TBM) of disordered systems using the scaling behavior of the typical density of states (GDOS) as localization criterion. The GDOS is obtained as the geometrical mean value of the local density of states (LDOS) averaged over many different lattice sites and disorder realizations. The LDOS can efficiently be obtained within the kernel polynomial method (KPM). To check the validity and accuracy of the method, we apply it here to the standard Anderson model of disordered systems, for which the results (for instance for the critical disorder strength of the Anderson transition) are well known from other methods.

Keywords: metal-insulator transition; typical density of states; local density of states; finite-size scaling; kernel polynomial method.

PACS numbers: 71.23.An, 71.30.+h, 72.20.Ee, 64.60.an, 71.23.-k

1. Model

We study the well known Anderson model of a disordered system[1],

$$\hat{H} = \sum_i \epsilon_i |i\rangle\langle i| + t \sum_{\substack{i,j \\ (NN)}} |j\rangle\langle i|. \qquad (1)$$

[*]present address

Here, i denotes the lattice sites of a 3-dimensional simple-cubic lattice, for which we consider a finite system consisting of $N = L^3$ sites with periodic boundary conditions; as usual, the site-diagonal matrix elements ϵ_i are considered to be (uncorrelated) random variables, drawn from a finite interval of width W (box distribution), and the fixed (non-random) hopping matrix element t is assumed to be non-vanishing for nearest neighbor sites only.

Though the localization properties of the Anderson model (critical disorder W_c, scaling behavior, etc.) are well known already from the literature[2], we consider it here once more to validate and test our numerical methods, which shall be applied to more realistic models in the future.

2. Local density of states and its arithmetic and geometric average

A physical quantity which is suitable to distinguish between localized and extended eigenstates is the local density of states (LDOS) at lattice site i,

$$\rho_i(E) = \sum_{k=1}^{N} |\langle i|k\rangle|^2 \delta(E - E_k), \quad (2)$$

where $|k\rangle$ denotes the eigenstate with eigenenergy E_k. It is intuitively clear that $\rho_i(E)$ is strongly fluctuating from site to site, if the states with eigenenergy near E are localized, as then only a few sites have a non-vanishing $\rho_i(E) > 0$. Therefore, to distinguish between energy regions of localized and delocalized states, one can calculate $\rho_i(E)$ for many different realizations of the disordered system (or, which should be equivalent, for different sites of a given realization) and calculate the arithmetic average of the local density of states (ADOS)

$$\rho_{\text{tot}}(E) = \frac{1}{S} \sum_{i=1}^{S} \rho_i(E), \quad (3)$$

which corresponds to the (averaged) total density of states, and the geometric average of the local density of states (GDOS)

$$\rho_{\text{typ}}(E) = \exp \frac{1}{S} \sum_{i=1}^{S} \log \rho_i(E). \quad (4)$$

which corresponds analogously to the so-called typical density of states[3]. S is the number of sites i considered (which may be further increased by studying different realizations of the random system).

In the thermodynamic limit $N \to \infty$ the GDOS approaches 0 for energies E in the region of (exponentially) localized states. Therefore, an investigation of the scaling behavior of the GDOS with increasing system size N (or L) can be used to distinguish between energy regions of localized and delocalized states and to determine the mobility edges[4] (ME) and the critical disorder strength W_c at which all states become localized (Anderson transition).

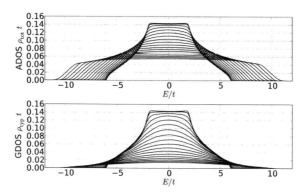

Fig. 1. The total (ADOS) and the typical (GDOS) density of states for different disorder parameters $W/t \in \{0, 1, 2, \ldots, 17\}$ (from top to bottom). System size is $N = 40^3 = 64000$ and truncation limit is $M = 140$ (see section 3).

The GDOS equals the ADOS for zero disorder. Otherwise it is, in general, smaller than the ADOS. Furthermore, it decreases with increasing disorder strength W (see fig. 1), system size N, or truncation limit M (see section 3).

3. The kernel polynomial method

Calculating the LDOS of tight-binding models (TBM) can most conveniently be done using the kernel polynomial method[5] (KPM), which is based on a polynomial series expansion using Chebychev polynomials,

$$f(x) = \frac{1}{\pi\sqrt{1-x^2}} \left(\mu_0 + 2\sum_{n=1}^{\infty} \mu_n T_n(x) \right). \tag{5}$$

Truncating this series after a finite number of moments M (truncation limit), the target function $f(x)$ may be approximated. The Chebychev polynomials are given by

$$T_n(x) = \cos(n \arccos(x)). \tag{6}$$

In the case of the LDOS at site i ($f(x) \equiv \rho_i(E)$), the coefficients μ_n (Chebychev moments) read

$$\mu_n^{(i)} = \int_{-1}^{1} f(x) T_n(x) \, dx = \langle i | T_n(H) | i \rangle. \tag{7}$$

Recurrence relations for the μ_n exist[5], which can easily be implemented numerically. The core of the iteration loop consists in a sparse matrix multiplication, hence memory consumption is quite low. Another advantage of the KPM is that, given an $N \times N$ sparse matrix representation of the hamiltonian $\hat{H}$ (1), the algorithm is an $O(N)$-method, i.e. it scales linearly with N, which means it shows good performance even for larger system sizes. By calculating the LDOS one avoids a complete diagonalization of H, which would scale as $O(N^3)$.

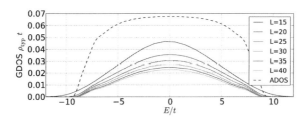

Fig. 2. Typical density of states (GDOS) for different system sizes N (disorder: $W/t = 13.2$).

4. Distinction between localized and extended states

As stated above, the scaling behavior of the GDOS as a function of the system size N (or system edge length L) shall be used as the localization criterion to decide in which energy region the states are localized or extended (i.e., to determine the MEs). For any finite system size, the GDOS also stays finite (even for localized states) and there is no sharp but only a gradual transition between extended and localized states. Fig. 2 shows the GDOS obtained for a fixed disorder strength ($W/t = 13.2$) and different system sizes $N = L^3$; for comparison, the ADOS is also shown. The GDOS decreases with increasing L and should approach 0 in the energy region of the localized eigenstates but a finite value in the region of delocalized states.

Therefore, the most simple ansatz for the scaling behavior of the GDOS is

$$\rho_{\text{typ}}(E, L) = \frac{a}{L^p} + b. \qquad (8)$$

Here, the parameters a and b and the exponent p must depend on the energy E. In the localized regime one has to expect $b = 0$ and a large exponent p. Fig. 3 shows the L-dependence of the GDOS for different energies E and again the disorder value $W/t = 13.2$.

If the ratio M/L^d ($d = 3$ for a 3D system) is kept constant for every investigated system size, it can be shown that directly at the transition the GDOS behaves like $1/L$, i.e. one can expect $p = 1$ exactly at the MEs. On the insulating side of the transition the scaling exponent increases ($p > 1$), and the GDOS value extrapolated for infinite system size is $b = 0$. On the metallic side, the exponent apparently stays roughly at $p = 1$, but the limit value is now finite ($b > 0$).

For convenience, we have assumed $b = 0$ also on the metallic side. It is clear that the data belonging to metallic states now cannot be fitted very well anymore. However, the fit algorithm still tries to find the best possible fit, resulting in curves that have an even smaller curvature. Thus, the scaling exponent takes values $p < 1$ in the metallic regime. This can be turned into an advantage, because now the MEs can easily be determined as those energies at which the exponent $p(E)$ crosses the value $p(E) = 1$ (see figure 4).

It is our goal to eventually find a fit model that fits the scaling behavior of the GDOS in all band areas (not just at the MIT) and to define a better localization criterion, but until then, the above procedure is already capable of determining the

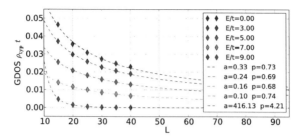

Fig. 3. Examples for the scaling behavior of the GDOS with increasing system edge length L for different band energies E (disorder: $W/t = 13.2$). Resulting fit parameters from fit model (8) in the legend.

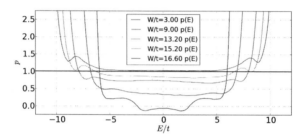

Fig. 4. Scaling exponent p as a function of energy E for different disorder parameters W, using the simple fit model with $b = 0$. The thick black line indicates the cutoff $c = 1$ that is used to determine the mobility edges.

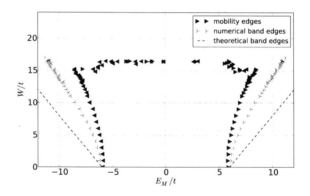

Fig. 5. Phase diagram of disorder as obtained by our method with the accuracy of our current data.

MIT for any range of disorder parameters W. This allows for the calculation of the phase diagram in the W-E plane, which shows the MEs E_M for different disorder strengths W and therefore the lines at which the MIT takes place (see figure 5).

Although the reached accuracy is still not satisfying – in particular close to the critical disorder strength $W_c/t \approx 16.5$ –, the phase diagram shown in figure 5 already features some well-known properties[3,5]:

- For zero disorder, the mobility edges coincide with the band edges.
- For increasing disorder, the energy spectrum broadens, and localized states appear in the band tails.
- Reentrance behavior: At a specific disorder parameter W_t, the width between the two MEs shrinks dramatically. However, the extremal energy values reached at these turning points are slightly different from the ones in the literature[3] ($E_t/t \approx 8.3$ instead of $E_t/t \approx 7.6$).
- The critical disorder is reached at about $W_c/t \approx 16.5$.

Conclusion

We have shown that even with the very crude and oversimplified ansatz (8) for the scaling law of the GDOS, good results for the MEs and the Anderson MIT and a reasonable estimate for the phase diagram can be obtained. However, to avoid fluctuations in the scaling exponent $p(E)$ (and thus in the position of the MEs), either the accuracy has to be further increased (especially near the critical disorder W_c), or a more sophisticated fit model has to be developed. Under the given circumstances, we are confident to be able to develop a method which does not rely on prior knowledge of external calibration parameters (e.g. the position of the critical disorder W_c) in the near future.

After successful implementation of the algorithms and application to the Anderson model (1), we want to apply the procedure to more interesting and more realistic systems, e.g. binary alloys and multi-band TBMs, describing, for instance, semiconductors with magnetic impurities.

References

1. P. W. Anderson, *Phys. Rev.* **109**, 1492 (1958).
2. B. Kramer, A. MacKinnon, *Rep. Prog. Phys.* **56**, 1469 (1993).
3. G. Schubert and H. Fehske, Quantum Percolation in Disordered Structures, in *Quantum and Semiclassical Percolation and Breakdown in Disordered Solids*, Lecture Notes in Physics, Vol. 762/2009 (Springer, Berlin/Heidelberg, 2009), p. 1–28.
4. N. Mott, *Physics Today* **31**, 42 (1978).
5. A. Weiße et al., *Rev. Mod. Phys.* **78**, 275–306 (2006).

CRITICAL EXPONENT FOR THE QUANTUM SPIN HALL TRANSITION IN $\mathbb{Z}_2$ NETWORK MODEL

K. KOBAYASHI

Department of Physics, Sophia University
Kioi-cho 7-1, Chiyoda-ku, Tokyo 102-8554, Japan
k-koji@sophia.ac.jp

T. OHTSUKI

Department of Physics, Sophia University
Kioi-cho 7-1, Chiyoda-ku, Tokyo 102-8554, Japan

K. SLEVIN

Department of Physics, Graduate School of Science, Osaka University
Machikaneyama 1-1, Toyonaka, Osaka 560-0043, Japan

We have estimated the critical exponent describing the divergence of the localization length at the metal-quantum spin Hall insulator transition. The critical exponent for the metal-ordinary insulator transition in quantum spin Hall systems is known to be consistent with that of topologically trivial symplectic systems. However, the precise estimation of the critical exponent for the metal-quantum spin Hall insulator transition proved to be problematic because of the existence, in this case, of edge states in the localized phase. We have overcome this difficulty by analyzing the second smallest positive Lyapunov exponent instead of the smallest positive Lyapunov exponent. We find a value for the critical exponent $\nu = 2.73 \pm 0.02$ that is consistent with that for topologically trivial symplectic systems.

Keywords: quantum spin Hall transition; critical exponent; network model.

PACS numbers: 71.30.+h, 73.20.Fz

1. Introduction

Onoda et al.[1] have questioned whether Anderson transitions in topologically non-trivial systems share the same critical properties as those in topologically trivial systems. Obuse et al.[2,3] studied the metal-ordinary insulator transition in quantum spin Hall (QSH) systems and found that the value of the critical exponent is the same as that of topologically trivial symplectic systems. However, Obuse et al. only studied transitions to insulating phases without edge states leaving open the possibility that the critical exponent for the metal-QSH insulator transition might be different. Indeed, the critical conductance distributions have been found to be sensitive to the number of the edge states.[4]

Here, we report an estimation of the critical exponent for the divergence of the localization length at the metal-QSH insulator transition. We find a value of the critical exponent that is consistent with that in the conventional symplectic class. Our result supports the conjecture that the metal-QSH insulator transition belongs to the conventional Wigner-Dyson symplectic class.

1.1. *Second smallest positive Lyapunov exponent*

In order to estimate the critical exponents for the divergence of the localization length, we have performed a finite size scaling analysis of numerical data for Lyapunov exponents.[5,6,7,8] This involves the estimation of the Lyapunov exponents for quasi-one-dimensional systems with effectively infinite length and finite cross sections. For extrapolation to the two dimensional limit that is of interest here, a strip with a cross section L is considered. For ordinary metal-insulator transitions, the standard approach is to analyze the scaling of the dimensionless quantity

$$\Gamma = \Gamma_1 = \gamma_1 L, \qquad (1)$$

which is equal to the product of the smallest positive Lyapunov exponent γ_1 (precisely defined below) and the cross section. (This quantity is the inverse of the so-called MacKinnon-Kramer parameter.) In a metallic phase Γ usually increases with L and in a localized phase usually decreases with L. At the transition point between these phases Γ becomes independent of L and a common crossing point of curves with different L is visible on the appropriate graph. However, when applying this method to the metal-QSH insulator transition we run into a problem: Γ increases in both the metallic and QSH insulating phases and no common crossing point is seen (Fig. 2). This problem, which makes a very precise scaling analysis difficult, occurs because of the existence of edge states in the QSH insulating phase. We have overcome this problem by analyzing the scaling of the quantity

$$\Gamma_2 = \gamma_2 L, \qquad (2)$$

which involves the second smallest positive Lyapunov exponent. This Lyapunov exponent is much less affected by these edge states and a common crossing point is recovered (Fig. 3). It has been demonstrated[9] for the Anderson transition in three dimensional systems in the orthogonal symmetry class that the critical exponent obtained from the scaling of higher Lyapunov exponents is the same as that obtained from the scaling of the smallest positive exponent.

2. Calculation of Lyapunov exponents

2.1. *Model*

To describe the QSH system, we use a $\mathbb{Z}_2$ quantum network model.[10,11] The $\mathbb{Z}_2$ network model has two controlling parameters, p, the tunneling probability (related to the chemical potential), and q, the spin-mixing probability (related to the spin-orbit interaction strength). For finite q, the $\mathbb{Z}_2$ network model exhibits a metallic

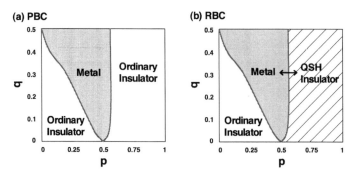

Fig. 1. Phase diagram of $\mathbb{Z}_2$ network model[10] with (a) PBC and (b) RBC. When RBC are imposed, the insulating phase without edge states appears for small p, the quantum spin Hall insulating phase appears for large p, and a metallic phase appears between the two insulating phases. The arrow indicates the range of the parameter considered in this paper.

phase sandwiched between two insulating phases. When periodic boundary conditions (PBC) are imposed, both insulating phases correspond to the ordinary insulating phase [Fig. 1(a)]. On the other hand, when reflecting boundary conditions (RBC) are imposed, the insulating phase located for larger p exhibits edge states and becomes the QSH insulating phase [Fig. 1(b)]. Thus, the $\mathbb{Z}_2$ network model with RBC shows two types of transition: the metal-ordinary insulator transition and the metal-QSH insulator transition.[a]

The cross section of the network is measured in terms of the number of links; a slice of width L contains L links, and each link has a spin degree of freedom. In this paper we focus on the case of L even with RBC. (For L odd, perfectly conducting channels appear.[12,13])

2.2. *Lyapunov exponents*

We consider a quasi-one-dimensional system. The Lyapunov exponents are estimated using a transfer matrix method. The transfer matrix T_x relates the current amplitudes ψ_x on the slice x to those on the next slice,

$$\psi_{x+1} = \boldsymbol{T}_x \psi_x. \qquad (3)$$

Consider the matrix $\mathcal{T}^\dagger \mathcal{T}$, where $\mathcal{T}$ is the transfer matrix product for the strip with length M,

$$\mathcal{T} = \boldsymbol{T}_M \cdots \boldsymbol{T}_2 \boldsymbol{T}_1. \qquad (4)$$

The matrix $\mathcal{T}^\dagger \mathcal{T}$ is Hermitian and positive definite (because $\mathcal{T}$ is invertible) and so its eigenvalues are real and positive. They also occur in reciprocal pairs because of

---

[a] The direct transition from the ordinary to QSH insulator occurs only for $q \to 0$. In this limit, the transition point belongs to the same universality class as the integer quantum Hall system.

current conservation. From each eigenvalue a Lyapunov exponent is defined by the limit

$$\gamma = \lim_{M \to \infty} \frac{\ln \lambda}{2M}, \tag{5}$$

where λ is an eigenvalue of $T^\dagger T$. In QSH systems, states are doubly degenerate (Kramers degeneracy) and the same degeneracy occurs in the Lyapunov exponents. Keeping this in mind, and putting the exponents in decreasing order, we number them as

$$\gamma_{L/2} > \gamma_{L/2-1} > \cdots > \gamma_2 > \gamma_1 > 0 > -\gamma_1 > -\gamma_2 > \cdots > -\gamma_{L/2}. \tag{6}$$

Here, it is to be understood that each exponent is doubly degenerate.

The calculation of the Lyapunov exponent Eq. (5) is terminated at finite M when the target Lyapunov exponent has converged to within a specified precision. Note that higher Lyapunov exponents converge more quickly than lower Lyapunov exponents.

3. Results

3.1. *Finite size scaling*

We consider a $\mathbb{Z}_2$ network model with strip geometry with RBC in transverse direction and the spin-mixing parameter set to $q = 0.309$ (where the transition point is known to be around $p = 0.561$[10]). For each Γ, we assume the scaling formula,

$$\Gamma(L, p) = F((p - p_c)L^{1/\nu}, (p - p_c)L^{y_{\text{irr}}}), \tag{7}$$

where $y_{\text{irr}}(< 0)$ is the exponent for the irrelevant correction to scaling. We expand the scaling function around the QSH transition point $p = p_c$ as

$$\Gamma(L, p) = \Gamma_c + \sum_{j=1}^{n_c} c_j \left(uL^{1/\nu}\right)^j + \sum_{i=1}^{2} \sum_{j=0}^{n_d(i)} d_j(i) \left(uL^{1/\nu}\right)^j (vL^{y_{\text{irr}}})^i, \tag{8}$$

where u is the relevant scaling variable,

$$u = \sum_{j=1}^{n_a} a_j \left(\frac{p - p_c}{p_c}\right)^j, \quad a_1 = 1, \tag{9}$$

and v the irrelevant scaling variable,

$$v = \sum_{j=0}^{n_b} b_j \left(\frac{p - p_c}{p_c}\right)^j, \quad b_0 = 1. \tag{10}$$

First, we analyze Γ, which is calculated from the smallest positive Lyapunov exponent (see Fig. 2). A fit of the numerical data to Eq. (8) with $n_a = 3$, $n_b = 1$, $n_c = 5$, $n_d(1) = 5$, $n_d(2) = 2$ and taking $p_c, \Gamma_c, a_j, b_j, c_j, d_j(i), \nu$, and y_{irr} as fitting parameters, yielded

$$p_c = 0.561 \pm 0.001, \quad \Gamma_c = 0.1400 \pm 0.0004, \quad \nu = 2.64 \pm 0.06, \quad y_{\text{irr}} = -0.94 \pm 0.04. \tag{11}$$

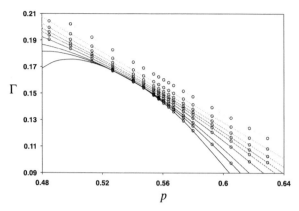

Fig. 2. Γ of Eq. (1) versus tunneling probability for the $\mathbb{Z}_2$ network model near the metal-QSH insulator transition. The lines are a finite size scaling fit, with different lines corresponding to different cross sections ($L = 16, 24, 32, 40, 48, 64, 96, 128, 192$, from top to down).

The precision of the data for the smallest positive Lyapunov exponents is better than 0.03% for $L = 16, 24, 32, 40, 48, 64$ and 0.05% for $L = 96, 128, 192$, and goodness of fit is 0.2. Nevertheless, the precision of the estimate of the critical exponent ν is not sufficient to distinguish it from the critical exponent of the quantum Hall transition, $\nu = 2.59 \pm 0.01$.[14]

Next we analyze Γ_2, which is calculated from the 2nd smallest positive Lyapunov exponent (see Fig. 3). We see a clear common crossing point of the curves separating the metallic and topological insulating phases. A fit of the numerical data to Eq. (7) with $n_a = 2$, $n_b = 1$, $n_c = 4$, $n_d(1) = 4$, $n_d(2) = 2$ yielded

$$p_c = 0.562 \pm 0.001, \ \Gamma_c = 1.429 \pm 0.004, \ \nu = 2.73 \pm 0.02, \ y_{\text{irr}} = -0.95 \pm 0.02. \quad (12)$$

The precision of the data for the 2nd Lyapunov exponents is better than 0.01%, and goodness of fit is 0.7. This result is in good agreement with that obtained by Asada et al.[15] for SU(2) model with PBC, $\nu = 2.746 \pm 0.009$, and by Obuse et al.[2] for $\mathbb{Z}_2$ network model with RBC at the metal-ordinary insulator transition, $\nu = 2.88 \pm 0.04$. It is also clearly different from the exponent for the quantum Hall transition.

4. Summary

We have reported an estimate of the critical exponent for the divergence of the localization length at the metal-quantum spin Hall insulator transition. By analyzing the scaling of the 2nd smallest positive Lyapunov exponent, we have estimated the critical exponent to be $\nu = 2.73 \pm 0.02$. Our result shows that this critical exponent is insensitive to the topological property of the insulating phase. This is in sharp contrast to the critical conductance distribution,[4] which is sensitive to

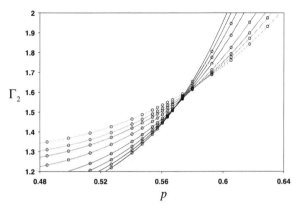

Fig. 3. Γ_2 of Eq. (2) versus tunneling probability for the $\mathbb{Z}_2$ network model near the metal-QSH insulator transition. The lines are a finite size scaling fit, with different lines corresponding to different cross sections ($L = 32, 40, 48, 64, 96, 128, 160, 192$, from flatter to steeper). The crossing point shifts slightly toward p_c with the increase of L.

the type of transition, *i.e.*, the presence or absence of edge states in the adjacent insulating phase.

Analysis of the scaling of higher Lyapunov exponents may also be useful in the study of other Anderson transitions where conducting edge or surface states occur in the insulating phase, such as, for example, in topological insulators.

Acknowledgments

This work was supported by KAKENHI No. 23·3743 and KAKENHI No. 23540376.

References

1. M. Onoda, Y. Avishai, and N. Nagaosa, Phys. Rev. Lett. **98**, 076802 (2007).
2. H. Obuse, A. Furusaki, S. Ryu, and C. Mudry, Phys. Rev. B **78**, 115301 (2008).
3. H. Obuse, A. R. Subramaniam, A. Furusaki, I. A. Gruzberg, and A. W. W. Ludwig, Phys. Rev. B **82**, 035309 (2010).
4. K. Kobayashi, T. Ohtsuki, H. Obuse, and K. Slevin, Phys. Rev. B **82**, 165301 (2010).
5. A. MacKinnon and B. Kramer, Phys. Rev. Lett. **47**, 1546 (1981).
6. A. MacKinnon and B. Kramer, Z. Phys. B **53**, 1 (1983).
7. B. Kramer and A. MacKinnon, Rep. Prog. Phys **56**, 1469 (1993).
8. B. Kramer, A. MacKinnon, T. Ohtsuki, and K. Slevin, Int. J. Mod. Phys. B **24**, 1841 (2010).
9. K. Slevin and T. Ohtsuki, Phys. Rev. B **63**, 045108 (2001).
10. H. Obuse, A. Furusaki, S. Ryu, and C. Mudry, Phys. Rev. B **76**, 075301 (2007).
11. S. Ryu, C. Mudry, H. Obuse, and A. Furusaki, New J. Phys. **12**, 065005 (2010).
12. T. Ando and H. Suzuura, J. Phys. Soc. Jpn. **71**, 2753 (2002).
13. K. Kobayashi, K. Hirose, H. Obuse, T. Ohtsuki, and K. Slevin, J. Phys.: Conf. Ser. **150**, 022041 (2009).
14. K. Slevin and T. Ohtsuki, Phys. Rev. B **80**, 041304(R) (2009).
15. Y. Asada, K. Slevin, and T. Ohtsuki, Phys. Rev. B **70**, 035115 (2004).

DISORDER INDUCED BCS-BEC CROSSOVER

AYAN KHAN

Research Center for Dielectric and Advanced Matter Physics,
Pusan National University, Busan, 609-735, South Korea
ayan.khan@pusan.ac.kr

Of late, the study of BCS-BEC crossover in the presence of weak random impurity is an interesting issue. In this proceedings we study the effect of this disorder which is included through the Nozières and Smith-Rink theory of superconducting fluctuations. In the weak regime, the random potential leaves an effect on the superconducting order parameter but it spares the chemical potential. Here we present the exact behavior of the mean field quantities as a function of the disorder by self-consistently solving the coupled equations.

Keywords: BCS-BEC Crossover; Disorder.

PACS numbers: 05.30.Jp, 74.20.Fg, 74.40.+k

1. Introduction

The smooth evolution of atoms from Cooper pairs to the composit bosons offers an exciting domain of research in current days. The exhilarating transformation of ultracold atomic gases can easily be viewed by changing inter-atomic low energy scattering length by means of Fano-Feshbach resonance[1]. This enabled us to study the Bardeen-Cooper-Schrieffer-Bose-Einstein-Condensate(BCS-BEC) crossover more closely[2,3].

The recent experimental developments widened the domain for more intriguing aspects in ultracold atomic gases, the effect of impurity driven disorder is among them. In the seminal work of Anderson[4], the localization effect due to impurity was predicted for the superconductors when impurity concentration is pretty high but this feature spares the material when it is less dirty. But a direct observation of Anderson localization in electronic systems is very difficult. On the contrary, ultracold atoms allow one to address the core of the phenomenon that Anderson had discovered, since they are genuine quantum particles described as matter waves, and interactions can be reduced at a negligible level so that one can study single particle behaviors. More over one can create disorder with optics and this lends a great amount of controllability and tunability in the system. Recently ultracold Bose gas in one dimension (^{87}Rb and ^{39}K) enabled us to see the localization directly[5,6]. Latest experiments are conducted in three dimension for both non interacting atomic Fermi gas of ^{40}K[7] and Bose gas of ^{87}Rb[8]. These experiments have widened the possibility to study the crossover in the light of disorder[9] experimentally.

In theoretical front, the static disorder in Fermi and in Bose systems are not a very new topic. A considerable amount of attention has been paid to disorder driven superconductors[10,11,12] and in Bose gas[13,14]. More detail references on disordered superconductors and Bose gas can be found in Ref.[9]. Of late the interest at unitarity is also gaining pace[15,16,17,18], but still it lacks the required amount of attention. Here, at present our investigation is on BCS-BEC crossover with weak uncorrelated disorder at zero temperature. To absorb the impurity in the mean field approach we follow Nozières and Smith-Rink (NSR) theory[19] of superconducting fluctuations. As a result we are able to construct the modified gap and density equation[15]. This enables us to solve the coupled equation self consistently as a function of the random external potential through out the crossover. At zero temperature which is not yet reported according to best of our knowledge.

We arrange our study in the following way, in Section 2 we present the formalism to include fluctuation in the mean field level. As mentioned above, very few article are available in this field at present therefore we prefer to spend some time here to discuss the intricacies. Section 3 is dedicated to our results and we draw our conclusion in Section 4.

2. Formalism

To describe the effect of impurity in Fermi superfluid from BCS to BEC regime one need to start from the real space Hamiltonian in three dimension for s-wave superfluid,

$$\mathcal{H}(\mathbf{x}) = \sum_\sigma \Psi_\sigma^\dagger(\mathbf{x}) \left[-\frac{\nabla^2}{2m} - \mu + V_d(\mathbf{x}) \right] \Psi_\sigma(\mathbf{x})$$
$$+ \int dx' V(\mathbf{x}, \mathbf{x}') \Psi_\uparrow^\dagger(\mathbf{x}') \Psi_\downarrow^\dagger(\mathbf{x}) \Psi_\downarrow(\mathbf{x}) \Psi_\uparrow(\mathbf{x}'),$$

where $\Psi_\sigma^\dagger(\mathbf{x})$ and $\Psi_\sigma(\mathbf{x})$ represents the creation and annihilation of fermions with mass m and spin state σ at $\mathbf{x}$ respectively, $V_d(\mathbf{x})$ signifies the random potential and μ is the chemical potential. All physical constants are are taken care using Plank units. In the interaction Hamiltonian the s-wave fermionic interaction is defined with $V(\mathbf{x}, \mathbf{x}') = -g\delta(\mathbf{x} - \mathbf{x}')$ and g is the bare coupling strength of fermion-fermion pairing. We choose the disorder as uncorrelated which reflects as white-noise correlations. By definition $V_d(\mathbf{x}) = \sum_i g_d \delta(\mathbf{x} - \mathbf{x}_i)$ where g_d is fermionic impurity coupling constant and $\mathbf{x}_i$ are the static positions of the quenched disorder. So the correlation function will turnout as $\langle V_d(-q) V_d(q) \rangle = \beta \delta_{i\omega_m, 0} \kappa$, where $q = (\mathbf{q}, i\omega_m)$. β is known as the inverse temperature, where as $\omega_m = 2\pi m/\beta$ is the bosonic Matsubara frequency with m being an integer. The disorder strength can be written as $\kappa = n_i g_d^2$, where n_i being the concentration of the impurities.

The partition function corresponding to the Hamiltonian can be written in the path integral formulation as

$$Z = \int \mathcal{D}[\bar{\Psi}, \Psi] \exp\left[-S(\{\bar{\Psi}\}\{\Psi\})\right], \qquad (1)$$

where the action $S = \int_0^\beta d\tau \int d\mathbf{x}[\bar{\Psi}_\sigma \partial_\tau \Psi_\sigma + H]$. An introduction of pairing field $\Delta(\mathbf{x}, \tau)$ and applying the Grassman identity ($\int \mathcal{D}[\bar{\Delta}, \Delta] \exp\left[-\int d\mathbf{x} \int_0^\beta d\tau \bar{\Delta}\Delta/g\right] = 1$) in Eq.(1) the new partition function can be written as,

$$Z = \int \mathcal{D}[\bar{\Psi}, \Psi] \int \mathcal{D}[\bar{\Delta}, \Delta] \exp\left[-S_{eff}\right]. \qquad (2)$$

The effective action, $S_{eff} = S(\bar{\Psi}, \Psi) + 1/g \int d\mathbf{x} \int_0^\beta d\tau \bar{\Delta}\Delta$. Following Hubbard-Stratonovich transformation, Eq.(2) can be written in terms of inverse Nambu propagator as,

$$Z_{eff} = \int \mathcal{D}[\bar{\Delta}, \Delta] e^{-1/g \int d\mathbf{x} \int_0^\beta d\tau \bar{\Delta}\Delta} \times \int \mathcal{D}[\bar{\Psi}\Psi] e^{-\int d\mathbf{x} \int_0^\beta d\tau \bar{\Psi} G^{-1} \Psi}, \qquad (3)$$

where the inverse Nambu propagator $G^{-1}(\mathbf{x}, \tau)$ is defined as, $G^{-1} = -\partial_\tau \mathbb{I} + (\nabla^2/2m + \mu - V_d)\sigma_z + \Delta\sigma_x$ where $\mathbb{I}$ the identity matrix and σ_i are the Pauli matrices ($i \in \{x, y, z\}$). So after integrating out the fermionic fields from Eq.(3) we are left with the effective action as,

$$S_{eff} = \int d\mathbf{x} \int_0^\beta d\tau \left[\frac{|\Delta(\mathbf{r})|}{g} - \frac{1}{\beta} \text{Tr} \ln\{-\beta G^{-1}(\mathbf{r})\}\right], \qquad (4)$$

where $\mathbf{r} = (\mathbf{x}, \tau)$. It is important to mention that the main contribution in the partition function comes from a small fluctuation $\delta\Delta(\mathbf{x}, \tau) = \Delta(\mathbf{x}, \tau) - \Delta$ where Δ is the homogeneous BCS pairing field. The original Green's function can be written as a sum of Green's function in absence of disorder ($G_0^{-1} = -\partial_\tau \mathbb{I} + (\nabla^2/2m + \mu)\sigma_z + \Delta\sigma_x$) and a self energy contribution ($\Sigma = -V_d\sigma_z + \delta\Delta\sigma_+ + \delta\bar{\Delta}\sigma_-$) which contains the disorder as well as the small fluctuations of the BCS pairing fields, where $\sigma_\pm$ are the ladder matrices.

It is possible to write the effective action (S_{eff}) in Eq.(4) as a sum of bosonic action (S_B) and fermionic action (S_F) by expanding the inverse Nambu propagator upto the second order. In other words, it suggests that the expansion of the self energy is carried out till the first order in the disorder strength (κ), which holds good as long as the disorder is weak. The effective action also contains a term, which emerges from the linear order of self energy expansion ($G_0\Sigma$). It is possible to set the linear order to zero if we consider S_F is an extremum of S_{eff} after performing all the fermionic Matsubara frequency sums. The constrained condition leads to the BCS gap equation which after appropriate regularization through s-wave scattering length reads,

$$-\frac{m}{4\pi a} = \sum_k \left[\frac{1}{2E_k} - \frac{1}{2\epsilon_k}\right]. \tag{5}$$

This suggests that the BCS gap equation does not have any contribution from the disorder potential. These conditions holds well when one considers the disorder as weak. Now to construct the density equation with usual prescription of statistical mechanics; the thermodynamic potential Ω should be differentiated with respect to the chemical potential μ. Ω can be written as a sum over fermionic (Ω_F) and bosonic (Ω_B) thermodynamic potentials. So density, $n = n_F + n_B = -\frac{\partial}{\partial \mu}(\Omega_F + \Omega_B) = -\frac{1}{\beta}\frac{\partial}{\partial \mu}(S_F + S_B)$. The well known BCS density equation can be restored if we consider only n_F but the presence of disorder and fluctuation leads to $n_B \neq 0$. Hence the final mean field density equation will be,

$$n = \sum_k \left(1 - \frac{\xi_k}{E_k}\right) - \frac{\partial \Omega_B}{\partial \mu}, \tag{6}$$

where Ω_B is defined as,

$$\Omega_B = \lim_{\beta \to \infty} \frac{1}{2\beta} \sum_q \ln \det M - \frac{\kappa}{2} \sum_{q,\omega_m=0} W^\dagger M^{-1} W. \tag{7}$$

The first term in Eq.(7) is the contribution arises due to the fluctuating pairing fields. This contribution becomes important if the objective of study relates to the finite temperature effect on the system. Since our purpose of study is the effect of disorder at zero temperature, from here on we will neglect this contribution. In Eq.(7), W is a doublet which couples the disorder with the fluctuation. At $T = 0$ after performing the fermionic Matsubara frequency summation one finds,

$$W_1 = W_2 = \sum_k \frac{\Delta(\xi_k + \xi_{k+q})}{2E_k E_{k+q}(E_k + E_{k+q})}. \tag{8}$$

The inverse fluctuation propagator is a 2×2 symmetric matrix and at zero temperature it reads,

$$M_{11} = \frac{1}{g} + \sum_k \left[\frac{v_k^2 v_{k+q}^2}{i\omega_m - E_k - E_{k+q}} - \frac{u_k^2 u_{k+q}^2}{i\omega_m + E_k + E_{k+q}}\right],$$

$$M_{12} = \sum_k u_k v_k u_{k+q} v_{k+q} \left[\frac{1}{i\omega_m + E_k + E_{k+q}} - \frac{1}{i\omega_m - E_k - E_{k+q}}\right], \tag{9}$$

with $M_{22}(q) = M_{11}(-q)$ and $M_{21}(q) = M_{12}(q)$. The other functions bear usual BCS notation with $\xi_k = k^2/2m - \mu$, $E_k = \sqrt{\xi_k^2 + \Delta^2}$, $u_k^2 = 1/2(1 + \xi_k/E_k)$ and $v_k^2 = 1/2(1 - \xi_k/E_k)$.

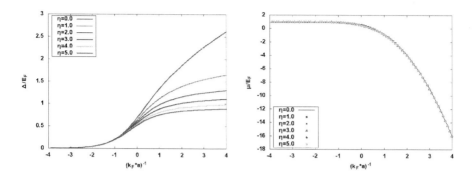

Fig. 1. The order parameter follows BCS mean field approximation of exponential decay in the BCS side, whereas in the BEC side it is more prone to the disorder and gets depleted due to destruction of superfluidity. Conversely the chemical potential remains unaffected.

Finally the disorder induced bosonic thermodynamic potential at zero temperature reads:

$$\Omega_B = \frac{\kappa}{2} \sum_{\mathbf{q}} \frac{2W_1^2}{M_{11} + M_{12}}, \qquad (10)$$

where the elements of the pair fluctuation matrix takes the form,

$$M_{11} + M_{12} = = \sum_k \left[\frac{\Delta^2 - E_k \xi_{k+q} - E_{k+q} \xi_k}{2 E_k E_{k+q}(E_k + E_{k+q})} + \frac{1}{2E_k} \right]. \qquad (11)$$

3. Result and Discussion

Eq.(5) and (6) are now ready to be solved self consistently, aided by the results of Eq.(8-11). It is clear from these equations that the disorder strength κ is an input parameter, this provokes the question about how to estimate the disorder strength as weak. Firstly κ has a dimension of k_F/m^2 which leads to the dimensionless disorder strength $\eta = \kappa m^2/k_F$. But a more physical description can be worked out if the impurity strength is normalized by Fermi density and square of Fermi energy[18]. According to the this new definition, $\tilde{\eta} = \kappa n_F/\epsilon_F^2 = 4/(3\pi^2)\eta$. Therefore it is essential that the impurity density and strength remain much less than Fermi density and Fermi energy.

Fig.1 demonstrates that in the BCS limit the bosonic contribution becomes nominal. As a result the order parameter with different disorder strength follows the mean field approximation of $\Delta/\epsilon_F = 8e^{-2} \exp[-\pi/(2k_F a)]$ corresponding to $1/k_F a \to -\infty$ thus emphasizing the validity of Anderson theorem[4]. In the BEC limit it is possible to make analytic extension for the bosonic thermodynamic potential by expanding the inverse fluctuation propagator matrix elements in powers of q^{20}. In effect the coefficients of q turns up simple analytic integrals. A systematic and careful calculation leads to an analytic description of the order parameter in this limit, where one can observe a progressive depletion. Here the impurity actually starts to destroy the superfluidity and in effect the order parameter is depleted and the fraction of depletion remains in the order of $\eta/(k_F a)$ in agreement with Ref [15]. Hence Anderson theorem breaks down in this limit. The chemical potential, expectantly remains pinned to the Fermi energy in the BCS limit and follows the ideal path (without disorder situation) towards the BEC limit.

The present study explicitly reveals that the disorder starts to play its role in the crossover window (which is usually considered as $1/k_F a \simeq [-1, 1]$) itself. It suggests, that the depletion of order parameter in vicinity of unitarity leads to the onset of superfluid destruction. Though there exist numerous discussions on the BCS and BEC extremes with disorder, but according to best of our knowledge, the behavior of the disorder induced order parameter is not yet been reported. Here for the first time we show this behavior from weak to strong coupling limit.

4. Conclusion

In conclusion, we included an impurity like weak disorder through the gaussian fluctuation route as prescribed earlier[15,18] and then solved the coupled BCS mean field equation self consistently. This enabled us to obtain the two mean field parameters Δ and μ where we show that the order parameter gets depleted leaving the chemical potential unchanged. These behaviors were commented in Ref.[15] but here we present the exact calculation for the first time. The gradual decrease of Δ can be connected to the destruction of superfluidity. It is already known[13] that the random potential destroys the superfluid nature in Bose gas hence the superfluid order parameter gets depleted. Further it reveals that the destructive process begins around the resonance point. We are now in process in understanding other aspects using this formalism[21]. In future it will be very interesting to study the crossover region under the influence of large impurity to fill the void of our knowledge in this situation. Therefore we hope a very exciting future for this fairly new topic in the coming days.

Acknowledgments

A.K would like to thank the Post Doctoral Grant from Research Center for Dielectric and Advanced Matter Physics, Pusan National University.

References

1. C. Chin, R. Grimm, P. Julienne, and E. Tiesinga, *Rev. Mod. Phys.* **82**, 1225 (2010).
2. S. Georgini, L. P. Pitaevskii, and S. Stringari, *Rev. Mod. Phys.* **80**, 1215 (2008).
3. I. Bloch, J. Dalibard, and W. Zwerger, *Rev. Mod. Phys.* **80**, 885 (2008).
4. P. W. Anderson, *J. Phys. Chem. Solids* **11**, 26 (1959).
5. J. Billy *et. al.*, *Nature* **453**, 891 (2008).
6. G. Roati *et. al.*, *Nature* **453**, 895 (2008).
7. S. S. Kondov, W. R. McGhee, J. J. Zirbel, and B. DeMarco, *Science* **334**, 66 (2011).

8. F. Jendrzejewski et. al., arXiv:1108.0137.
9. L. S. Palencia, and M. Lewenstein, Nature Phys. **6**, 87 (2010).
10. D. Belitz, and T. R. Kirkpatrick, Rev. Mod. Phys. **66**, 261 (1994).
11. A. Ghoshal, M. Randeria and N. Trivedi, Phys. Rev. B **65**, 014501 (2001).
12. K. Bouadim, Y. L. Loh, M. Randeria, and Nandini Trivedi, Nature Phys. **7**, 884 (2011).
13. K. Huang, and H. F. Meng, Phys. Rev. Lett. **69**, 644 (1992).
14. S. Giorgini, L. Pitaevskii, and S. Stringari, Phys. Rev. B **49**, 12938 (1994).
15. G. Orso, Phys. Rev. Lett. **99**, 250402 (2007).
16. B. I. Shklovskii, Superconductors **42**, 909 (2008).
17. C. A. R. Sá de Melo, Physics Today **61**, 45 (2008).
18. L. Han, and C. A. R. Sá de Melo, New J. Phys. **13**, 055012 (2011).
19. P. Nozières S. Smith-Rink, J. Low Temp. Phys. **59**, 195 (1985).
20. E. Taylor, A. Griffin, N. Fukushima, and Y. Ohashi, Phys. Rev. A **74**, 063626 (2006).
21. A. Khan, S. Basu and S. W. Kim, arXiv:1202.5117.

A COMPARISON OF HARMONIC CONFINEMENT AND DISORDER IN INDUCING LOCALIZATION EFFECTS IN A SUPERCONDUCTOR

POULUMI DEY[1], AYAN KHAN[2], SAURABH BASU[1,3] and B. TANATAR[3]

[1]*Department of Physics, IIT Guwahati, Guwahati, Assam 781039, India*
[2]*Research Center for Dielectric and Advanced Matter Physics, Pusan National University, Busan, 609-735, South Korea*
[3]*Department of Physics, Bilkent University, Bilkent, 06800 Ankara, Turkey*

We present a comparative study of the localization effects induced by harmonic trapping and random onsite disorder in a 2D s-wave superconductor. Performing a numeric computation of the Bogoliubov-de Gennes equations in the presence of parabolic trap and random disorder, we obtain the eigensolutions as a function of trap depth and disorder strength. While the wavefunctions demonstrate localization tendencies for moderate disorder strengths, with finally yielding an insulating behavior at large disorder, the harmonic confinement effects are seen to be vastly more drastic in inducing localization effects.

Keywords: BCS-BEC Crossover; Disorder.

PACS numbers: 74.20.Fg

1. Introduction

The interplay between localization and superconductivity is an old problem. The gradual vanishing of superconducting transition temperature as a function of the driving parameter and finally the emergence of the insulating phase signaled via enhanced resistivity represents a classic example for the onset of an insulating phase. In many of the thin films, disorder plays the role of the driving parameter which at large values of disorder show considerable digression from early theories of dirty superconductors[1,2]. The issue of superfluid-insulator transition (SIT) received renewed attention with the observation of inhomogeneous pairing with formation of isolated superconducting islands in a highly disordered s-wave superconductors[3,4]. At the mean-field level, the eigenstates of the system become localized and in the limit of large disorder, the superconducting regions reduce to small 'blobs' which are separated by extended insulating strips. At extremely large disorder, superconductivity vanishes giving way to a homogeneous insulating phase.

Similar effects can also be produced by external confining potentials. Creating such confinement effects for charge carriers on a crystal lattice is an impossible task as the wavelength of the electromagnetic wave that is needed to create such a trap should be of the order of the lattice spacing. However, the situation for atomic gases (bosons as well as fermions) in optical lattices present a far more optimistic scenario to achieve such trapping effects[5,6]. Similar signatures of transition to an insulating phase has been observed as a function of strength of the harmonic trapping potential[7].

Motivated by the above ideas, we examine the transition from a superconductor to an insulator state in a 2D superconductor described by a negative-U Hubbard model that includes disorder or trapping effects. The disorder is modeled by a gaussian distribution characterized by its full width at half maximum (FWHM), σ and the trapping potential is assumed to have a harmonic form with a minimum at the center of the lattice, which gradually becomes shallow at the edges and is characterized by a single parameter V_0 (> 0) (see discussion in section 2).

In this work, the signature of the transition from a superfluid to an insulator is seen via a non-monotonic behavior of the spectral gap. Initially it shows a decrease with increasing perturbation for both kinds of inhomogeneity, but beyond a certain value, the spectral gap starts to increase. The order parameter that characterizes the superfluid phase however vanishes in both cases with further enhancement of the perturbation term. The two results collectively indicate onset of a gapped phase that has no long range order. Additional support for the localization effects is provided by a drastic drop in the participation ratio that confirms the localization of the trapped eigenstates to individual lattice sites.

We organize our paper as follows. In order to fix notations, we provide a brief description of our model and the key equations. In the next section, we present results for the spectral gap which shows a transition to a *localized* insulating phase with increasing strength of the trapping potential. Similar features are also noted for local density calculation where mobility of the charge carriers reduces with increasing interaction energy. A rapid decay in the participation ratio with increasing disorder and trapping strengths, provide support for such a state. We conclude with a brief summary of our results.

2. Model

We consider the two-dimensional Hubbard model in which the atoms interact attractively via the on-site interaction, U and a random on-site disorder or a harmonic trapping potential, V_i,

$$\mathcal{H} = -t \sum_{\langle ij \rangle, \sigma} (c_{i\sigma}^\dagger c_{j\sigma} + \text{H.c.}) - |U| \sum_i \left(n_{i\uparrow} - \frac{1}{2}\right)\left(n_{i\downarrow} - \frac{1}{2}\right) + \sum_{i,\sigma} (V_i - \mu) n_{i\sigma} \quad (1)$$

All the symbols have usual meaning. V_i is modeled by $P[V_i(\sigma)] = \frac{1}{\sqrt{2\pi\sigma^2}} e^{-V_i^2/2\sigma^2}$ and $V_i = V_0(r_i - r_0)^2$ for disorder and harmonic trapping, respectively. σ and V_0 are strengths of the disorder and trapping potential and r_0 is the position where the center of the trap lies which is located at the center of the lattice in our case. Thus the potential is minimum (deepest) at the center of the lattice and is maximum (shallow) at the edges. All of U, μ, σ and V_0 are expressed in units of the hopping strength, t.

The effective Hamiltonian from Eq. (1) can be rewritten as,

$$H_{eff} = \sum_{ij\sigma} H_{ij\sigma}(c_{i\sigma}^\dagger c_{j\sigma} + \text{H.C}) + \sum_i (\Delta_i c_{i\uparrow}^\dagger c_{i\downarrow}^\dagger - \Delta_i^* c_{i\uparrow} c_{i\downarrow}), \qquad (2)$$

where, $H_{ij\sigma} = -t\delta_{i\pm1,j} - (\mu + |U|\langle n_i\rangle/2 - V_i)\delta_{ij}$ and $\langle n_i\rangle = \sum_\sigma \langle n_{i\sigma}\rangle$. Eq. (2) is solved using standard Bogoliubov-de Gennes (BdG) mean-field method. Different steps of the BdG calculations[8] are given elsewhere, thus are not repeated here for brevity. We only quote the important equations that are subsequently numerically solved on a two-dimensional lattice.

The gap parameter and density, to be solved self-consistently, in terms of the eigenvectors $u_n(\mathbf{r}_i)$ and $v_n(\mathbf{r}_i)$ at temperature T are given by

$$\Delta_i = -|U|\sum_n [u_n(\mathbf{r}_i)v_n^*(\mathbf{r}_i)f(E_{n\uparrow}) - u_n(\mathbf{r}_i)v_n^*(\mathbf{r}_i)f(-E_{n\downarrow})], \qquad (3)$$

$$\langle n_{i\sigma}\rangle = \sum_n [|u_n(\mathbf{r}_i)|^2 f(E_{n\sigma}) + |v_n(\mathbf{r}_i)|^2 f(-E_{n\bar\sigma})] \qquad (4)$$

where $f(E_{n\sigma})$ is the Fermi distribution function. Δ_i and $\langle n_{i\sigma}\rangle$ are obtained self-consistently from Eq. (3) and Eq. (4) at each lattice site. In this work, we specialize to at zero temperature, hence the Fermi function becomes unit step-function. The dimension of the lattice is chosen to be 24×24 with periodic boundary conditions for the disorder case and with open boundary conditions for the trapped case.

We have worked with a moderate U, i.e. $U = 2.5t$, so as to mimic the BCS limit and density is kept at two-third filling ($n = 0.66$). Our results are qualitatively insensitive to the choice of densities, which we have verified in our numerics. Densities lower than $n < 0.1$ are avoided as they yield vanishingly small superconducting correlations. The physical quantities corresponding to the disordered case are averaged over sufficient number of disorder realizations.

3. Results and Discussion

Before we proceed with the issue of SIT, we wish to brief on the nature of the superconducting state which subsequently undergoes a transition to an insulating state at large disorder potential and trap depths. To comment on the effects of the perturbation on the superconducting state, we present the local gap parameter, Δ_i by solving Eq. (3) and Eq. (4) self-consistently. For the disordered case, it may be seen that there are formations of *islands* as the strength of disorder is increased from a low ($\sigma = 0.5t$) to a large value ($\sigma = 3t$) where Δ_i's are finite (Fig. 1). The superconducting islands shrink in size and are separated by large insulating strips characterized by $\Delta_i \equiv 0$ in the limit of large disorder. While for the trapped case, as the strength of the confinement potential is increased, Δ_i's acquire strength near the potential minima, as seen by the concentric circles of the contour plots in Fig. 2,

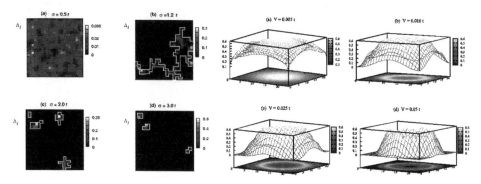

Fig. 1. Δ_i in the presence of a random on-site disorder for $\sigma = 0.5t, 1.2t, 2t$ and $3t$. The non-zero Δ_i regions are marked to show the formation of superconducting islands.

Fig. 2. Δ_i in the presence of a harmonic trapping potentials $V_0 = 0.005t, 0.016t, 0.025t$ and $0.05t$.

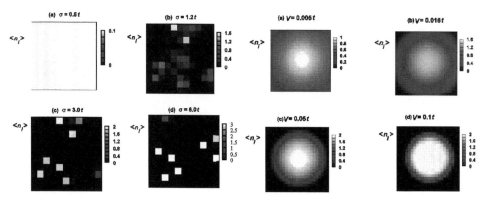

Fig. 3. $\langle n_i \rangle$ for the disordered case. A large σ, viz. $\sigma = 6t$ is included to show the complete localization of density at some particular sites.

Fig. 4. $\langle n_i \rangle$ in the presence of trapping potentials. Again a large V_0, viz. $V_0 = 0.1t$ shows pairs ($n = 2$) occupying the center of the trap, leaving very few carriers at the periphery.

as the confinement is tuned from a very low ($V_0 = 0.005t$) to a moderately low value ($V_0 = 0.05t$).

The plots for the local densities, $\langle n_i \rangle$ in Figs. 3 and 4 corroborate the above observations. The superconducting islands again characterize the case of random disorder, while the trapped case demonstrates a distinct rise in the occupancy of sites near the trap center with increasing V_0. Along with the inferences from Fig. 2, it is apparent that the system separates into a phase that comprises of localized bound pairs at the core, surrounded by conducting ring of excess unpaired carriers. The constant occupation at large trap depths bear the signatures of a localized insulator as also noted in Ref. [11, 12].

An important quantity which provides an estimate of how superconductivity is affected is the spectral gap, E_{gap} which is the difference between the ground state

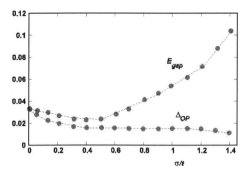
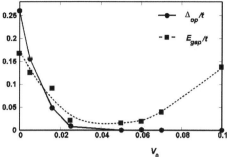

Fig. 5. E_{gap} and Δ_{op} are shown as a function of σ. Note that E_{gap} depends on σ in a non-monotonic fashion while Δ_{op} vanishes monotonically as σ is increased.

Fig. 6. E_{gap} and Δ_{op} are shown as a function of V_0. The behavior of E_{gap} and Δ_{op} with V_0 is similar to that of σ, however the effect is much more drastic. Δ_{op} vanishes for a small V_0.

and the lowest lying state of the excitation spectrum obtained from the BdG eigenvalues. E_{gap}, thus obtained is plotted in Figs. 5 and 6 as functions of increasing disorder and trapping strengths along with the superconducting order parameter, Δ_{op} defined by the long distance behavior of the correlation $\langle c_{i\uparrow}^{\dagger} c_{i\downarrow}^{\dagger} c_{j\downarrow} c_{j\uparrow} \rangle \to \Delta_{op}^{2}/|U|^{2}$ for $|\mathbf{r}_i - \mathbf{r}_j| \to \infty$[10]. While E_{gap} shows a non-monotonic behavior with increasing disorder and trapping strengths, Δ_{op} rapidly drops to zero. A finite E_{gap} along with a vanishing Δ_{op} for larger values of the perturbation potentials clearly point towards the emergence of an insulating state in the limit of large confinement[3].

Table 1. Participation ratio, PR is shown for various values of σ and V_0. Note that it reduces drastically with increasing σ and V_0 and approaches unity at the limit of extreme limit, albeit showing a more gradual behavior for the disordered case.

σ	PR	V_0	PR
0.0	575	0.0	498
0.25	313	0.001	127
0.75	43	0.005	22
1.0	24	0.016	11
1.2	5	0.025	8
1.5	4	0.06	3
3.0	1	0.1	1

Further, we analyze the behavior of the participation ratio (PR) as a function of confining potential in order to strengthen our claim of the localization scenario. PR gives a measure of the number of lattice sites over which an eigenstate is extended[13]. It distinguishes localized states (PR $\sim$ 1) from the extended ones (PR $\sim$ N, where N is total number of lattice sites) and hence can be used to predict localization of atomic states. The PR is defined as, $\text{PR}(E) = 1/\sum_{r_i} |\phi_E(r_i)|^4$, where $\phi_E(r_i)$ is the eigenstate obtained from the numerical solution of the BdG Hamiltonian (Eq. (2)) and E is the corresponding eigenvalue. In particular PR is computed corresponding to the ground state (i.e $E = E_0$) as a function of increasing disorder and trapping strengths (see Table 1). It may be noted that PR is of the order of a few lattice sites for low trapping strengths which makes superconductivity still feasible as the atomic wavefunction spreads out over some finite number of sites and long range order persists (see the marked regions in Fig. 1). But as σ and V_0 are made stronger, PRs drop to a value equal to one, implying the eigenstate to be confined at one (or a few) individual site(s). The localization of the eigenstates in the limit of extreme confinement, is also evident from the reduction in the localization length, ξ. Suppose an electronic wavefunction is localized within a d-dimensional volume of average diameter ξ which can be roughly taken as a characteristic length for the asymptotic exponential decay, called the localization length[14], the PR behaves as ξ^d. In our $d = 2$ case an estimate of the correlation length can be made as, $\xi = \sqrt{PR}$. The data for ξ reflects the same behavior as that of PR, hence not repeated here.

4. Conclusions

In this paper, we compared the localization effects brought about by random on-site disorder and a harmonic confinement for a s-wave superconductor. The onset of the localized phase as seen via a number of physically realizable observables is very efficiently induced by the trapping potential while the evolution towards a localized phase is slower in the disordered case.

Acknowledgments

We thank DST, India for financial support through Grant No.-03(1097)/07/EMR-II and SR/S2/CMP-23/2009. A. K. like to thank the Post Doctoral Fellowship at Pusan National University. B. T. is supported by TUBITAK (109T267) and TUBA.

References

1. P.W. Anderson, *J. Phys. Chem. Solids* **11**, 26 (1959).
2. A. A. Abrikosov and L. P. Gorkov, *Sov. Phys. JETP* **9**, 220 (1959).
3. A. Ghosal, M. Randeria and N. Trivedi, *Phys. Rev. Lett.* **81**, 3940 (1998); *ibid Phys. Rev. B* **63**, 020505(R) (2000); *ibid Phys. Rev. B* **65**, 014501 (2001).
4. P. Dey and S. Basu, *J. Phys.: Condens. Matter* **20**, 485205 (2008).
5. M. Greiner, O. Mandel, T. Esslinger, T. W. Hansch and I. Bloch, *Nature* **415**, 39 (2002). *Phys. Rev. B* **40**, 546 (1989).
6. R. Jördens, N. Strohmaier, K. Günter, H. Moritz and T. Esslinger, *Nature* **455**, 204 (2008).
7. X. Liu, P.D. Drummond and H. Hu, *Phys. Rev. Lett.* **94**, 136406 (2005).
8. P.G. de Gennes, *Superconductivity in Metals and Alloys* Ben- jamin, New York, (1966).
9. Q. Cui and K. Yang, *Phys. Rev. B* **78**, 054501 (2008).
10. J. R. Waldram, *Superconductivity of Metals and Cuprates* Institute of Physics Publishing, London, (1996).
11. A.E. Feiguin and F. Heidrich-Meisner, *Phys. Rev. B* **76**, 220508(R) (2007).
12. N. Gemelke, X. Zhang, C. Hung and C. Chin, *Nature* **460**, 995 (2009).
13. F. Wegner, *Z. Physik. B* **36**, 209 (1980).
14. B. Kramer, *Springer Ser. Solid state Sciences* **83**, 138 (1998).

QUASI TWO-DIMENSIONAL NUCLEON SUPERFLUIDITY UNDER LOCALIZATION WITH PION CONDENSATION

TATSUYUKI TAKATSUKA

Faculty of Humanities and Social Sciences, Iwate University
Morioka 020-8550, Japan
takatuka@iwate-u.ac.jp

The aim here is to show an example for localization and relevant low-dimensional superfluids in nuclear system. Due to a particular property of tensor force originating from the One-Pion-Exchange (OPE) between two nucleons, dense nuclear medium undergoes a layer confinement of the nucleons on one hand and also pion condensation (PC) for pion field mediating two-nucleon interaction on the other hand. The localization is characterized by a layered structure with a specific spin-isospin ordering. In that situation, the pairing problem has a two-dimensional (2D) character, i.e., low-dimensional superfluid realized in the hadronic matter with strong interactions. The pairing description suitable to the 2D nature is presented and possible realization of superfluidity in neutron stars is discussed, togeter with its effect on the cooling scenarios.

Keywords: Localization; Superfluidity; Neutron stars.

1. Tensor Force and Localization

Historically, study on localization in nucleon matter was initiated by a motivation to explain giant glitches of Vela pulsar through a starquake[1] (corequake[2]) made possible if the NS has a solid core. Realistic calculations@[3,4], however, have shown that the short-range repulsion of two-nucleon interaction V (1, 2) is not so steep as to realize the quantum solid through the geometrical caging mechanism, as compared to the He system. In this context, it is of particular interest to ask whether another solidification mechanism is existent or not in nuclear medium. The answer is 'yes'[5–7]. We pay attention to the noncentral character of V (1, 2) which consists of the central, spin-orbit and tensor components;

$$V(1,2) = V_C(r) + V_{LS}(r)\mathbf{L}\cdot\mathbf{S} + V_T(r)S_{12}, \qquad (1)$$

$$S_{12} = 3(\boldsymbol{\sigma}_1\cdot\hat{\mathbf{r}})(\boldsymbol{\sigma}_2\cdot\hat{\mathbf{r}}) - (\boldsymbol{\sigma}_1\cdot\boldsymbol{\sigma}_2), \qquad (2)$$

with S_{12} the tensor operator depending on the spin ($\boldsymbol{\sigma}$) and direction ($\hat{\mathbf{r}}$). For a parallel-spin pair, we have the diagonal spin matrix element $<S_{12}> = (3cos^2\theta - 1)$ and see that for the isotriplet state of a nn or pp pair (the isosinglet state of a np pair), the $\theta \simeq \pi/2(\theta \simeq 0)$ direction is energetically most favorable relative position

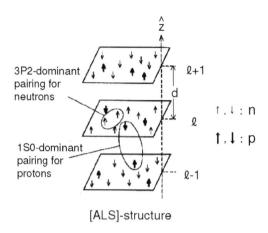

Fig. 1. ALS structure with one-dimensional localization and specific spin-isospin ordering.

because of $V_T > 0 (V_T < 0)$. Here θ is the angle between the spin quantization axis (z-direction) and direction $\hat{\mathbf{r}}$ with $\mathbf{r} = \mathbf{r}_1 - \mathbf{r}_2$. Similarly for an antiparallel-spin pair, $<S_{12}> = (-3cos^2\theta + 1)$ means that the $\theta \simeq 0$ ($\theta \simeq \pi/2$) is most favorable for a nn or pp pair (np pair).

We remark the points; (i) In usual Fermi gas (FG) phase, the first-order effects from tensor force vanish as a whole. (ii) In order to utilize the tensor force effects attractively, the specific spin-isospin ordering and at the same time, the localization of nucleons are necessary. (iii) Thus we have a new solidification (localization) mechanism if the nuclear system with (ii) is energetically favorable compared to FG phase.

2. ALS Model and Pion Condensation

From a viewpoint of large kinetic energy (KE) increase due to localization ($3\hbar\omega/4 >$ KE of 3D-FG), the one-dimensional (1D) localization, instead of usual 3D one, is advantageous because of smaller zero-point energy and is found not to lose the energy gain from interaction. Thus we propose an interesting structure of nucleon matter illustrated in Fig.1, called Alternating-Layer-Spin (ALS) structure taking account of the tensor force attraction most efficiently[6]. It is characterized by a layered structure due to 1D-localization and by a specific spin-isospin ordering, i.e., in the ℓ-th layer, the spin of like nucleons are aligned in the combination (n↑, p↓) and the spin directions change alternately layer by layer.

In the ALS model, the single-particle state is given by the following orthogonal basis functions $\{\phi_\alpha(\xi)\}$;

$$\phi_\alpha(\boldsymbol{\xi}) \equiv \phi(\boldsymbol{\xi}; \mathbf{q}_\perp, \ell) = exp(i\mathbf{q}_\perp \mathbf{r}_\perp)/\sqrt{\Omega_\perp} \cdot \phi_\ell(z)\chi_\alpha(\text{spin, isospin}), \qquad (3)$$

where $\boldsymbol{\xi} \equiv (\mathbf{r}, \text{spin, isospin})$, $\mathbf{q}_\perp \equiv (q_x, q_y)$, $\mathbf{r}_\perp \equiv (x, y)$, $\Omega_\perp$ is the 2D-normalization volume and the suffix α of the spin-isospin function χ means the combination of

(σ_ℓ, τ_ℓ) as $\sigma_\ell \tau_\ell = -(-1)^\ell$, $\sigma_\ell(n) = (-1)^\ell$ and $\sigma_\ell(p) = (-1)^{\ell+1}$ with $\tau_\ell = +1(-1)$ for p(n). $\phi_\ell(z)$ is given by the Bloch or Wannier function and is well approximated by a Gaussian function, $\phi_\ell(z) = (a/\pi)^{1/4} exp[-a(z-d\ell)^2/2]$ for a developed localization, with d the layer spacing.

How about the pion field ϕ in the ALS structure? Due to the 1D localization and spin-isospin ordering, we have a periodic spin-isospin density wave which provides a static source function $\mathbf{S}(\mathbf{r})$ in z-direction and thereby ϕ has a non-zero expectation value $(<\phi> \neq 0)$ through the field equation for the $\pi - NP$-wave interaction ($H_{\pi N}$ with $\boldsymbol{\sigma} \cdot \boldsymbol{\nabla}$-coupling),

$$(\boldsymbol{\nabla}^2 - m_\pi^2) < \phi >= -(f/m_\pi)\boldsymbol{\nabla}\mathbf{S}(\mathbf{r}). \tag{4}$$

The non-vanishing $<\phi> \neq 0$ means that pions participate as a real component of the medium owing to a particular structure of nucleon system, in contrast to $<\phi>=0$ for the usual FG state. This situation is nothing but the pion condensation (PC) which has gathered much attention since the first proposal by A.B. Migdal[8] and independently by R.F. Sawyer and D.J. Scalapino[9]. The condensed pion field (in the present case, neutral pion field $<\phi_0>$ as seen from Eq. (4)) has a form of a standing wave mode $<\phi_0> \propto \sin k_0 z$ with the condensed momentum $k_0 = \pi/d$. The equivalence between ALS structure and neutral pion (π^0) condensation has been shown through the equivalence between field description and potential description[6,7,10].

The ALS structure (i.e., π^0 condensation) is realized when the condensation energy (attractive contribution from OPE tensor) overwhelms the kinetic energy increase due to localization ($=\hbar\omega/4$+KE of 2D-FG−KE of 3D-FG). The onset density $\rho_t(\pi^0)$ is estimated as $\rho_t(\pi^0) \simeq (2-4)\rho_0$, with ρ_0 (=0.17 nucleons/fm^3 $\simeq 2.8 \times 10^{14}$ g/cc) being the nuclear density.

So far we have concentrated on the π^0 condensation. When we pay our attention to the isospin degrees of freedom of the pion field, we have energetically more favorable structure by allowing the coexistence of π^0- and π^c (charged pion)-condensations[7]. We extend the ALS model in such a way as the π^c condensation of a running wave mode is realized within the ALS layer ($x-y$ plane) with the condensed momentum $\mathbf{k}_c$ in $\mathbf{r}_\perp$ direction perpendicular to $\mathbf{k}_0$ in the z-direction. In this case, the nucleon basis function is given by a superposition of n- and p-ALS wave functions and the ALS structure corresponding to the combined neutral and charged pion condensations ($\pi^0 \pi^c$ condensation) is given likely in Fig.1 with a replacement of n by η, called 'new neutron'($\mid \eta >= u \mid n > +v \mid p >$, $u^2+v^2=1$) including n as a dominant component. A remarkable point is that both condensations are realized without serious interference and the energy gains from them are additive. Thus the ground state for nucleon matter with PC is given by the extended ALS model with $\pi^0 \pi^c$ condensation. By a realistic treatment, the onset density $\rho_t(\pi^0\pi^c)$ has been estimated as $\rho_t(\pi^0\pi^c) \simeq (2.5-4)\rho_0$[12].

3. Low-Dimensional Superfluidity

According to the increase of ρ, the nuclear lattice in the crust phase of NSs tends to resolve and at $\rho \gtrsim 0.5\rho_0$ the matter is in a uniform system, called 'liquid core', primarily composed of n as a main component and p by several % and also leptons (e^-, μ^-) assuring the charge neutrality. In nucleon matter, the most attractive pair state responsible for a realization of superfluidity depends on the fractional density $\rho_i (i = n, p)$. Here a pair state is specified by a set of quntum numbers $\tilde{\lambda} \equiv (T, S, L, J)$ with T, S, L and J being the pair's isospin, spin, orbital and total angular momenta, respectively, due to characteristics of nuclear force $V(1,2)$ in Eq. (1). In the case of pp pair near the Fermi surface, the scattering energy $E_{NN}^{LAB} (= 4\times$ Fermi energy$=4\epsilon_{iF} \simeq 244(\rho_i/\rho_0)^{2/3}$, $i = n, p$) is low because of small contamination ($\rho_p << \rho$) and the $^1S_0$ pair state, most attractive at low E_{NN}^{LAB}, should be responsible for superfluidity. On the other hand, the $^3P_2$ pair state, most attractive at high E_{NN}^{LAB}, should be responsible for neutron superfluidity, dominance in components ($\rho_n \simeq \rho$) and high E_{NN}^{LAB}. That is, the $^1S_0$-superfluidity for p and the $^3P_2$-one for n are expected. The former is a well known type as a S-state superfluidity, but the latter is a new one. This possibility of the $^3P_2$-superfluid was firstly pointed out by R. Tamagaki[13] and also by M. Hoffberg et al.[14], which was prior to the discovery of P-state superfluid of ^3H_e system in 1972. Realistic investigations by R. Tamagaki and myself[15-17] led to a conclusion that the n $^3P_2$-superfluid is realized in the density region $\rho \simeq (1-4)\rho_0$ of NSs.

Now we go to the superfluid problem under ALS structure. Since the above existence-region of the $^3P_2$ superfluid overlaps that of the ALS phase, our main concern is whether the $^3P_2$-superfluidity persists or not in the ALS localization. Due to 1D localization, the Fermi surface turns out to be of cylindrical shape and the single-particle energy is expressed as $\epsilon(q_\perp) = \hbar^2 q_\perp^2 / 2m_N^* +$const. with m_N^* the effective mass. The pairing correlation for the $(\mathbf{q}_\perp \ell, -\mathbf{q}_\perp \ell f)$-Cooper pair is operative in the $\mathbf{r}_\perp$-space where the 2D Fermi gas nature holds. To see what pair state is most attractive, the 2D plane wave of the pair wave function is expanded as[18]

$$e^{i\mathbf{q}_\perp \mathbf{r}_\perp} = \sum_{m_L} (i)^{m_L} J_{m_L}(q_\perp r_\perp) e^{im_L \varphi_r} e^{-im_\perp \varphi_q}, \qquad (5)$$

where partial waves are specified by m_L (the 2D angular momentum). Localization makes the pairing interaction in the same layer predominant and for simplicity an approximation to ignore the contribution from the pairs with $\ell \neq \ell'$ is taken. Then the spin part of the wave function is characterized by $S = 1$ and $m_S = (-)^\ell$ (z-component of S) due to the spin alignment. Finally, a pair state in the ALS scheme is specified by a set of quantum numbers $\tilde{\lambda} = \{S = 1, m_S = (-)^\ell, m_L\}$ instead of λ of 3D case.

What $\tilde{\lambda}$-pair is most attractive? We construct the 2D effective potential $\tilde{V}_{\tilde{\lambda}}(r_\perp)$ by taking average of $V(1, 2)$ over the localized wave $\phi_{rel}(z) \equiv (a/2\pi)^{1/4} \exp(-az^2/4)$

as

$$V_{\tilde{\lambda}}(r_\perp) \equiv <\tilde{\lambda} \mid V(1,2) \mid \tilde{\lambda}> = \tilde{V}_c(r_\perp) + m_S m_L \tilde{V}_{LS}(r_\perp) + \tilde{V}_T(r_\perp). \quad (6)$$

Eq.(6) indicates that $V_{\tilde{\lambda}}(r_\perp)$ is more attractive for the pair with $m_L, m_S = m_L(-)^\ell > 0$ because of $\tilde{V}_{LS}(r_\perp) < 0$. Consequently we should take the combination; m_L=1, 3,... for $m_S = 1$ (ℓ=even layers) and m_L=-1, -3, ... for $m_S = -1$ (ℓ=odd layers), with the restriction $\mid m_L \mid$=1 due to antisymmetrization. Among possible m_L, $\mid m_L \mid$=1 is most effective due to the centrifugal effects. The state with $\tilde{\lambda} \equiv \{S = 1, m_S = m_L = 1 \text{ or } m_S = m_L = -1\}$ contains the contribution from the $^3P_2$-interaction as a main contributor and is the most attractive pair state in the $\tilde{\lambda}$ scheme at high densities. This pairing is suitably called '$^3P_2$-dominant pairing'[18]. The energy gap equation for the $^3P_2$-dominant pairing has a 2D character in the points; 2D interaction in Eq. (6), 2D single-particle energy $\epsilon(q_\perp)$ and the 2D momentum space. Numerical calculations of the energy gap show that the $^3P_2$-superfluidity in usual 3D case can persist as a $^3P_2$-dominant superfluidity of 2D character, even if the 1D ALS-localization comes into play in NS cores. By the way, the proton $^1S_0$-dominant superfluidity (superconductivity) can be treated quite similarly.

However this conclusion is for a simple case where pure nn-pair and ALS-phase only with π^0 condensation are taken into account. More realistic discussions need to include the following points; (i)As has been discussed in the preceding section, the ground state with PC should be described by the extended ALS with a coexistence of π^0 and π^c condensations. (ii) Isobar Δ effects important for the realization of ALS. The point (i) requires that the pairing correlation has to be treated on the basis of η-particles. In this case, the attraction for the $^3P_2$-dominant pairing is weakened due to the fact that the η-state is a superposition of n and p states, which is called 'attenuation effect'and works against the occurrence of superfluidity. For the point (ii), the η-particle state is extended as a quasi-baryon state given by a superposition of nucleon and isobar states and the BCS state due to the $^3P_2$-dominant pairing is treated on this quasi-baryon baisis. In this case, the attraction is enhanced mainly through the enhancement of $\tilde{V}_T(r_\perp)$ in Eq.(6), working for the realization of superfluidity. As a net effects from (i) and (ii), the $^3P_2$-dominant superfluidity of quasibaryons is found to occur in NS cores[19], with the critical temperature $T_c \sim (0.5 - 2) \times 10^9$ K exceeding the internal temperature $T_{in} \simeq 10^8$ K of NSs.

4. Concluding Remarks

We want to stress that a layered structure of nucleon matter, a solid-like state of 1D localization with a specific spin-isospin ordering, is possible due to the characteristics of nuclear force (tensor force), which is nothing but the pion-condensed state in the field description. The system is well described by the ALS model. It is remarked that under the situation the pairing correlation becomes of a 2D character, presenting an interesting example of low-dimensional superfluid in hadronic matter with strong interactions.

Finally we want to stress the importance of superfluidity in a context of NS cooling. Observed colder class NSs(such as Vela, Geminga, 3C58 and so on) cannot be explained by a standard cooling process (modified URCA) but encounter the problem of 'too rapid cooling'when rapid cooling process (extremely efficient ν-emission by usual β-decay type in exotic phases) is applied, that is, require the rapid cooling and the superfluidity to suppress too rapid cooling, at the same time. In this sense, the so-called 'Pion cooling'(β-decay of η-particles) is a promising candidate, since both requirements above are satisfied by the ALS-phase with PC. It is wortwhile to note that this pion cooling necessarily shows up as a unique candidate, because another promising one, the so-called 'hyperon cooling'(hyperon β-decay in hyperon-mixed NS cores), though successfully applied, is faced to a serious difficulty as to the occurence of Λ-superfluidity[20].

Acknowledgments

The author expresses a special thank to R. Tamagaki for continuous collaboration on the series of work presented here. He also thanks S. Tsuruta, T. Tatsumi, T. Muto, S. Nishizaki and J. Hiura for useful discussions and encouragements.

References

1. G. Baym, C.J. Pethick, D. Pines and M. Ruderman, *Nature* **224**, 872 (1969).
2. D. Pines, J. Shaham and M. Ruderman, *Nature Phys. Sci.* **237**, 83 (1972).
3. D.M. Ceperley, G.V. Chester and M.H. Kalos, *Phys. Rev.* D**13**, 3208 (1976).
4. D.N. Lowy and C-. W. Woo, *Phys. Rev.* D**13**, 3201 (1976).
5. T. Takatsuka, K. Tamiya and R. Tamagaki, *Prog. Theor. Phys.* **56**, 685 (1976).
6. T. Takatsuka, K Tamiya, T. Tatsumi and R. Tamagaki, Prog. *Theor. Phys.* **59**, 1933 (1978).
7. T. Takatsuka, R. Tamagaki and T. Tatsumi, *Prog. Theor. Phys. Suppl.* No. **112** ,67 (1993).
8. A.B. Migdal, Zh. *Exsper. Theor. Fiz.* **63**, 1993 (1972); *Sov. Phys. JETP* **36**, 1052 (1973).
9. R.F. Sawyer and D.J. Scalapino, *Phys. Rev.* D**7**, 953 (1972).
10. T. Takatsuka and J. Hiura, *Prog. Theor. Phys.* **60**, 1234 (1978).
11. T. Kunihiro, T. Takatsuka, R. Tamagaki and T. Tatsumi, *Prog. Theor. Phys. Suppl.* No. **112**, 123 (1993).
12. T. Muto, R. Tamagaki and T. Tatsumi, *Prog. Theor. Phys. Suppl.* No. **112**, 159 (1993).
13. R. Tamagaki, *Prog. Theor. Phys.* **44**, 905 (1970).
14. M. Hoffberg, A.E. Glassgold, R.W. Richardson and M. Ruderman, *Phys. Rev. Lett.* **24**, 775 (1970).
15. T. Takatsuka and R. Tamagaki, *Prog. Theor. Phys.* **46**, 114 (1971).
16. T. Takatsuka, *Prog. Theor. Phys.* **48**, 1517 (1972).
17. T. Takatsuka and R. Tamagaki, *Prog. Theor. Phys.* **112**, 37 (2004).
18. T. Takatsuka and R. Tamagaki, *Prog. Theor. Phys. Suppl.* No. **112**, 107 (1993).
19. R. Tamagaki and T. Takatsuka, *Prog. Theor. Phys.* **116**, 573 (2006); **117**, 861 (2007).
20. S. Tsuruta, J. Sadino, A. Kobeiski, M.A. Teter, A.C. Liebmann, T. Takatsuka, K. Nomoto and H. Umeda, *Ap. J.* **691**, 691 (2009).

Localisation 2011
International Journal of Modern Physics: Conference Series
Vol. 11 (2012) 139–144
© World Scientific Publishing Company
DOI: 10.1142/S2010194512006034

ENHANCEMENT OF GRAPHENE BINDING ENERGY BY Ti 1ML INTERCALATION BETWEEN GRAPHENE AND METAL SURFACES

TOMOAKI KANEKO*,† and HIROSHI IMAMURA*,‡,§

*Nanoscale Theory Group, NRI, AIST
1-1-1 Umezono, Tsukuba, Ibaraki 305-8565, Japan
§h-imamura@aist.go.jp

Received 31 December 2011
Revised 20 March 2012

We theoretically investigate the effects of intercalation of a thin Ti layer between graphene (Gr) and metal surface (Au, Ag, Al, Pt, and Pd) on structural and electronic properties by using the first-principles total energy calculations. We find that the strongest binding energy is realized when 1 monolayer (ML) of Ti is intercalated between Gr and a metal surface independent of the metal atoms, which is 0.08-0.15 eV larger than that for Gr absorbed on a Ti(0001) surface. As the number of Ti layers increases, the binding energy monotonically decreases and converges to that for Gr adsorbed onto the Ti surface. We show that the origin of the enhancement of binding energy can be classified into two classes by considering the affinity of Ti for the metal surfaces.

Keywords: Graphene; First-principles calculations; Binding energy

PACS numbers: 71.15.-m, 73.22.Pr, 71.15.Nc, 61.72.up, 73.29.At

1. Introduction

Graphene (Gr) is an atomically thin carbon sheet that has attracted considerable interest for its application to future electronic devices by virtue of its outstanding transport properties due to the Dirac cone dispersion.[1] To utilize Gr as a basic element of future electronic devices, we need to study the physics of the interface between Gr and metal electrodes. A Ti/Au bilayer is widely used as a metal electrode for Gr[2,3,4,5,6,7] because Gr is tightly adsorbed onto Ti surfaces compared with other metal surfaces. Recently we have investigated the electronic properties of Gr on Ti/Au surfaces using the first-principles calculations.[8] We reported that the Gr binding energy is 24% increased compared with that of Ti surface when Ti 1ML is intercalated between Gr and an Au surface.[8] However, in that report we did not

---

†First-Principles Simulation Group, CMSU, NIMS, 1-2-1, Sengen, Tsukuba, Ibaraki 305–0047, Japan
‡Spintronics Research Center, AIST, 1-1-1 Umezono, Tsukuba, Ibaraki 305-8565, Japan

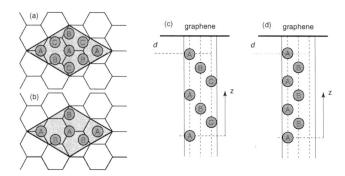

Fig. 1. (color online) Schematic illustration of the model of Gr structure on M surface. This figure is modified from Ref. [8]. Panels (a) and (b) show top views of Gr on fcc (111) and hcp (0001) M surfaces, respectively. Panels (c) and (d) show side views of Gr on fcc (111) and hcp (0001) M surfaces, respectively.

discuss in detail the enhancement of binding energy. The purpose of the present study is to investigate how Ti 1ML intercalation increases the Gr binding energy.

The effect of contacting the M electrode on Gr is a fundamental issue for the Gr-based device application and many authors have investigated this issue theoretically.[9,10,11] The strength of Gr binding on the M surface has been regarded as an important factor, because strong binding is needed in order to prevent Gr exfoliation from the M surface. According to the extensive studies by Giovannetti, Khomyakov, and their co-workers, Gr is physisorbed onto Au, Ag, Al, Pt, and Cu surfaces, while chemisorbed onto Pd, Ti, Ni and Co surfaces.[9,10] The Dirac cone is preserved for physisorbed Gr but disturbed for chemisorbed Gr.[9,10,11] Thin Ti layer intercalation between Gr and M has been introduced in the electrodes that measure electronic transport in order to tighten Gr binding.[2,3,4,5,6,7,12,13]

In this paper, we theoretically investigate the effects of Ti intercalation between Gr and several M (Au, Ag, Al, Pt, and Pd) surfaces using the first-principles total energy calculations. We show that the Gr binding energy takes the largest value when 1 ML of Ti is intercalated between the Gr and M surfaces independent of the M species. We elucidate how Ti 1ML intercalation increases the Gr binding energy, and we classify the origin of the enhancement of the Gr binding energy into two classes by considering the affinity of Ti for the M surfaces.

2. Computational details

We performed the first-principles calculations based on the density functional theory within the projector-augmented wave[14] with the plane-wave basis set and the local density approximation (LDA)[15] implemented on VASP code.[16,17] Gr is absorbed onto a single side of a metallic slab surface consisting of seven layers as shown in Fig. 1, which is modified from Ref. [8]. In Figs. 1 (a) and (c), the structures of Gr absorbed onto M surface with fcc (111) structure are schematically illustrated. This model is used for Gr adsorbed onto Au, Ag, Al, Pt, and Pd surfaces and Ti intercalated surfaces. For Gr absorbed onto the Ti (0001) surface, we used the structure shown in Figs. 1 (b) and (d). The cell size is measured by Gr lattice constant $a = 2.446$ Å which is optimized value of free standing Gr in the LDA. Then, the lateral size of the 2×2 supercell of Gr matches the $\sqrt{3} \times \sqrt{3}$ structures of the M surfaces commensurately with mismatch $-1.7 \sim 2.2$ % for optimized bulk lattice structure in the LDA. In the cell, we introduced a vacuum region approximately 14 Å thick and used dipole correction[18] to exclude spurious dipole interactions between periodic images. We relaxed the positions of the C atoms and M atoms in the topmost three layers, but in the topmost four layers in the case of Ti 3ML intercalation. The cut-off energy of the plane wave basis set and the total energy difference were chosen to be 400 eV and 10^{-6} eV, respectively. We used k-point sampling $18 \times 18 \times 1$ for Gr adsorbed onto Au(111), Ag(111), Al(111), Pt(111), Pd(111), and Ti intercalated surfaces, $12 \times 12 \times 1$ for Ti(0001) surfaces, and comparably dense k-point sampling for corresponding pristine M surfaces.

3. Results and Discussion

Table 1 summarizes the calculated results of the Gr binding energy, $\Delta E^{(\mathrm{Gr})}$, the averaged separation between Gr and the topmost M atoms, d, the work function of Gr adsorbed onto the M surface, $W_{\mathrm{Gr/M}}$, and the work function of the pristine M surface, W_{M}. The binding energy of Gr is defined as

$$\Delta E^{(\mathrm{Gr})} = \frac{4E_{\mathrm{Gr}} + 3E_{\mathrm{M}} - E_{\mathrm{Gr/M}}}{8}, \quad (1)$$

Table 1. Calculated results of the binding energy of Gr per C atom, $\Delta E^{(\mathrm{Gr})}$, the separation of Gr between the topmost M atom, d, the work function with Gr, $W_{\mathrm{Gr/M}}$, and the work function of the pristine M surface, W_{M}.

	$\Delta E^{(\mathrm{Gr})}$ (eV)	d (Å)	$W_{\mathrm{Gr/M}}$ (eV)	W_{M} (eV)
Au(111)	0.034	3.37	4.79	5.63
Ti 1ML/Au	0.470	2.05	3.91	4.67
Ti 2ML/Au	0.355	2.06	3.97	4.72
Ti 3ML/Au	0.379	2.05	3.87	4.76
Ag(111)	0.032	3.33	4.23	4.90
Ti 1ML/Ag	0.459	2.05	3.90	4.64
Ti 2ML/Ag	0.346	2.06	3.98	4.70
Ti 3ML/Ag	0.378	2.05	3.89	4.77
Al(111)	0.028	3.48	4.06	4.20
Ti 1ML/Al	0.473	2.04	3.91	4.87
Ti 2ML/Al	0.359	2.06	3.95	4.85
Ti 3ML/Al	0.372	2.05	3.90	4.76
Pt(111)	0.045	3.26	4.93	6.00
Ti 1ML/Pt	0.527	2.04	4.10	4.60
Ti 2ML/Pt	0.362	2.07	3.89	4.71
Ti 3ML/Pt	0.374	2.05	3.89	4.79
Pd(111)	0.070	2.45	4.19	4.59
Ti 1ML/Pd	0.508	2.05	4.04	4.55
Ti 2ML/Pd	0.361	2.06	3.91	4.70
Ti 3ML/Pd	0.379	2.05	3.89	4.82
Ti(0001)	0.377	2.05	3.87	4.71

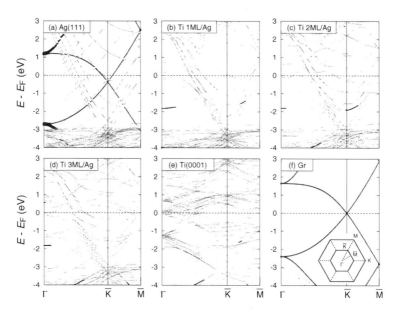

Fig. 2. (color online) The band structures of Gr adsorbed onto Ag(111) and Ti intercalated Ag surfaces. In these Figs. the projections to the p_z orbital of the C atoms are indicated by the blue dots, whose radii are proportional to the magnitude of projection.

where E_{Gr}, E_{M} and $E_{\text{Gr/M}}$ are the total energies of freestanding Gr, the pristine M surface and Gr absorbed onto M surface, respectively. We found that Gr is physisorbed onto Au(111), Ag(111), Al(111), and Pt(111) surfaces, but chemisorbed onto Pd(111), Ti(0001), and Ti/M surfaces.[9,10,11] For all M surfaces, as shown in Table 1, the Gr binding energy takes its maximum value when Ti 1ML is intercalated between Gr and M surfaces.[8] As the thickness of the Ti layer increases, the Gr binding energy monotonically decreases and rapidly converges to that of the Ti surface. d, $W_{\text{Gr/M}}$ and W_{M} also rapidly saturate to that of the Ti surface. Due to the larger resistivity of Ti than other M, it is desirable that an intercalated Ti layer be as thin as possible. Therefore, a Ti 1 ML intercalated surfaces is a good candidate for a practical M electrode for Gr devices.

Figure 2 shows the band structures of Gr adsorbed onto Ag and Ti intercalated Ag surfaces. The band structures of Gr adsorbed onto Ti (0001) and Gr in the doubled unit cell, respectively, are shown in Figs. 2 (e) and (f) as the references. The corresponding Brillouin zone is presented in the inset of Fig. 2 (f). In these Figs. the projections to the p_z orbital of the C atoms are indicated by the blue dots, whose radii are proportional to the magnitude of projection. For Gr on adsorbed onto Ag(111) surface, the Dirac cone survived and the Dirac point is shifted to -0.382 eV due to electron doping from the Ag surface to Gr. However, the Dirac cone disappears by the Ti 1ML intercalation between Gr and Ag surface because the p_z orbital of the C atoms strongly hybridize with d orbital of the Ti atoms. For other M surfaces, the Dirac cone also disappears even by Ti 1ML intercalation.[8]

Table 2. Calculated binding energy of Ti 1ML, $\Delta E^{(\text{Ti1ML})}$, and Gr/Ti 1ML, $\Delta E^{(\text{Gr/Ti1ML})}$, enhancement of Gr binding energy, $\Delta E_M^{(\text{Gr})} - \Delta E_{\text{Ti}}^{(\text{Gr})}$, and binding energy of Ti adatom, $\Delta E^{(\text{Ti})}$.

M	$\Delta E^{(\text{Gr/Ti1ML})}$ (eV)	$\Delta E^{(\text{Ti1ML})}$ (eV)	$\Delta E_M^{(\text{Gr})} - \Delta E_{\text{Ti}}^{(\text{Gr})}$ (eV)	$\Delta E^{(\text{Ti})}$ (eV)
Au(111)	4.61	1.89	0.09	4.42
Ag(111)	4.22	1.53	0.08	3.39
Al(111)	4.56	1.83	0.10	3.87
Pt(111)	5.75	2.87	0.15	7.86
Pd(111)	5.43	2.61	0.13	7.58
Ti(0001)	4.72	2.25	—	5.94

Finally we consider how Ti 1ML intercalation increases the Gr binding energy. The enhancement of Gr binding energy by Ti 1ML intercalation from that of the Ti surfaces can be written as follows:

$$\Delta E_{\text{Ti1ML/M}}^{(\text{Gr})} - \Delta E_{\text{Ti}}^{(\text{Gr})} = \frac{3}{8}(\Delta E_M^{(\text{Gr/Ti1ML})} - \Delta E_{\text{Ti}}^{(\text{Gr/Ti1ML})} - \Delta E_M^{(\text{Ti1ML})} + \Delta E_{\text{Ti}}^{(\text{Ti1ML})}), \quad (2)$$

where $\Delta E^{(\text{Ti1ML})}$ and $\Delta E^{(\text{Gr/Ti1ML})}$ are the binding energies of Ti 1ML and Gr/Ti 1ML per Ti atom, respectively. The calculated $\Delta E^{(\text{Ti1ML})}$ and $\Delta E^{(\text{Gr/Ti1ML})}$ are summarized in Table 3.

For the Au(111), Ag(111) and Al(111) surfaces, the binding energies of Gr/Ti 1ML and Ti 1ML are smaller than those for the Ti(0001) surfaces. The differences in the binding energies of Ti 1ML on these M surfaces between Ti(0001) are much larger than those of Gr/Ti 1ML. Therefore, the enhancement of the Gr binding energy is caused by the weak binding of Ti 1ML to these M surfaces. On the other hand, both Gr/Ti 1ML and Ti 1ML are tightly adsorbed onto Pt(111) and Pd(111) surfaces compared with those onto Ti(0001) surfaces. In these cases, the tight adsorption of Gr/Ti 1ML onto the M surfaces enhances the Gr binding energy by Ti 1ML intercalation. We found that the enhancement of the Gr binding energy can be classified into two classes.

This classification is based on the affinity of the Ti atom for M surfaces. The last row of Table 2 presents the binding energy of Ti adatom on the $\sqrt{3} \times \sqrt{3}$ structure M surface per Ti atom $\Delta E^{(\text{Ti})}$. We found that adsorption of Ti atom on Au(111), Ag(111), and Al(111) surfaces are much weaker than that on Ti surface, while those on Pd(111) and Pt(111) surfaces are much stronger than that on Ti surface. Since these behaviors agree well with those for the binding of Ti 1ML, these results indicate that the different affinities of Ti and several M surfaces activate different mechanisms of the enhancement of Gr binding energies.

4. Summary

We investigated the effects of the intercalation of a thin Ti layer between graphene and several metal surfaces (Au, Ag, Al, Pt ,and Pd) on structural and electronic properties using the first-principles total energy calculations. We found that Ti 1ML intercalation enhances the graphene binding energy by 0.08-0.15 eV from that for the

Ti surface. The graphene binding energy monotonically decreases as the thickness of the intercalated Ti layer increases. The enhancement of the binding energy can be divided into two classes. For Au, Ag, and Al surfaces, the enhancement of graphene binding energy is caused by the weaker binding of Ti 1ML on M surfaces than on Ti surfaces. For Pt and Pd surfaces, on the other hand, the enhancement of graphene binding energy is caused by the stronger binding of the bilayer of graphene and Ti 1ML than the Ti surface. The origin of the difference is the different affinities of Ti and different metal surfaces.

Acknowledgments

The authors acknowledge T. Nakano, Y. Shimoi, K. Harigaya, and H. Arai for their valuable discussions. This research was supported by the International Joint Work Program of Daeduck Innopolis under the Ministry of Knowledge Economy (MKE) of the Korean Government.

References

1. K. S. Novoselov, A. K. Geim, S. V. Morozov, D. Jiang, Y. Zhang, S. V. Dubonos, I. V. Grigorieva and A. A. Firsov, Science, **306**, 666 (2004).
2. E. J. H. Lee, K. Balasubramanian, R. T. Weitz, M. Burghard and K. Kern, Nature Nanotechnology, **8**, 486 (2008).
3. B. Huard, N. Stander, J. A. Sulpizio and D. Goldhaber-Gordon, Phys. Rev. B, **78**, 121402(R) (2008).
4. K. Nagashio, T. Nishimura, K. Kita and A. Toriumi, Appl. Phys. Lett. **97**, 143514 (2010).
5. S. Russo and M. F. Craciun, M. Yamamoto, A. F. Morpurgo and S. Tarucha, Physica E, **42**, 677 (2009).
6. J. A. Robinson, M. LaBella, M. Zhu, M. Hollander, R. Kasurda, Z. Hughes, K. Trumbull, R. Cavalero and D. Snyder, Appl. Phys. Lett. **98**, 053103 (2011).
7. P. Blake, R. Yang, S. V. Morozov, L. A. Ponomarenko, A. A. Zhukov, R. R. Nair, I. V. Grigorieva, K. S. Novoselov, A. K. Geim, Solid State Commun., **149**, 1068 (2009).
8. T. Kaneko and H. Imamura, Appl. Phys. Lett., **98**, 261905 (2011).
9. G. Giovannetti, P. A. Khomyakov, G. Brocks, V. M. Karpan, J. van den Brink and P. J. Kelly, Phys. Rev. Lett., **101**, 026803 (2008).
10. P. A. Khomyakov, G. Giovannetti, P. C. Rusu, G. Brocks, J. van den Brink and P. J. Kelly, Phys. Rev. B, **79**, 195425 (2008).
11. C. Gong, G. Lee, B. Shan, E. M. Vogel, R. M. Wallance and K. Cho, J. Appl. Phys., **108**, 123711 (2010).
12. H. B. Heersche, P. Jarillo-Herrero, J. B. Oostinga, L. M. K. Vandersypen and A. F. Morpurgo, Nature, **446**, 56 (2007).
13. T. Mueller, F. Xia, M. Freitag, J. Tsang, and Ph. Avouris, Phys. Rev. B, **79**, 245430 (2009).
14. P. Blöchl, Phys. Rev. B, **50**, 17953 (1994).
15. J. P. Perdew and A. Zunger, Phys. Rev. B, **23**, 5048 (1981).
16. G. Kresse and J. Furthmuller, Phys. Rev. B, **54**, 11169 (1996).
17. G. Kresse and J. Furthmuller, Comp. Mat. Sci., **6**, 15 (1996).
18. J. Neugebauer and M. Scheffler, Phys. Rev. B, **46**, 16067 (1992).

GENERALIZATION OF CHIRAL SYMMETRY FOR TILTED DIRAC CONES

TOHRU KAWARABAYASHI

Department of Physics, Toho University, Miyama 2-2-1
Funabashi, 274-8510, Japan
tkawa@ph.sci.toho-u.ac.jp

YASUHIRO HATSUGAI

Institute of Physics, University of Tsukuba,
Tsukuba, 305-8571 Japan

TAKAHIRO MORIMOTO and HIDEO AOKI

Department of Physics, University of Tokyo,
Hongo, Tokyo, 113-0033 Japan

The notion of chiral symmetry for the conventional Dirac cone is generalized to include the tilted Dirac cones, where the generalized chiral operator turns out to be non-hermitian. It is shown that the generalized chiral symmetry generically protects the zero modes ($n = 0$ Landau level) of the Dirac cone even when tilted. The present generalized symmetry is equivalent to the condition that the Dirac Hamiltonian is elliptic as a differential operator, which provides an explicit relevance to the index theorem.

Keywords: massless Dirac fermions; chiral symmetry; zero modes.

PACS numbers: 73.43.-f, 11.30.Rd, 73.61.Ph

1. Introduction

The chiral symmetry plays an important role in specifying some of the universality classes of the critical phenomena in disordered systems.[1–3] Systems are called chiral-symmetric when there exists an operator Γ that anti-commutes with the Hamiltonian, $\{\Gamma, H\} = 0$, with $\Gamma^\dagger = \Gamma$ and $\Gamma^2 = 1$. With this symmetry the energy eigenvalues appear always in pairs $(E, -E)$, since if we have an eigenstate ψ_E with an eigenvalue E, the state $\Gamma\psi_E$ is an eigenstate with an eigenvalue $-E$. The energy spectrum is therefore exactly particle-hole symmetric even when there exists a disorder as far as the disorder respects the chiral symmetry. In particular, the zero-energy state can be expressed as an eigenstate of Γ.

For a massless Dirac fermions in two dimensions, usually with vertical and isotropic Dirac cones as in graphene,[4–6] the effective Hamiltonian is expressed as $H_0 = v_F(\sigma_x \pi_x + \sigma_y \pi_y)$, where $\sigma_{x(y)}$ is the Pauli matrix and $\boldsymbol{\pi} = \boldsymbol{p} + e\boldsymbol{A}$ denotes the dynamical momentum with the vector potential $\boldsymbol{A}$ and the electron charge $-e$. The fermi velocity is denoted by v_F. For such a vertical Dirac cone, we have obviously $\{H_0, \sigma_z\} = 0$ and thus the system is chiral-symmetric with $\Gamma = \sigma_z$. The zero-energy Landau level (zero modes), which is essential to the anomalous quantum Hall effect for massless Dirac fermions in a magnetic field, then becomes an eigenstate of σ_z. The most remarkable property with these zero modes is their robustness against disorder in gauge degrees of freedom. The zero energy ($n = 0$) Landau level does not acquire any width due to such a disorder, while other Landau levels are broadened as usual, and this gives rise to an unconventional criticality for the quantum Hall transition at the $n = 0$ Landau level.[7] This robustness of zero modes for a vertical Dirac cone has been discussed in terms of the index theorem[8,9] or based on the explicit form of wave functions due to Aharonov and Casher.[10,11] It has been also demonstrated numerically that the chiral symmetry is also crucial to this robustness of zero modes of vertical Dirac cones.[12]

Massless Dirac fermions in two dimensions appear not only in graphene but also in certain organic metals, where we encounter pairs of tilted Dirac cones.[13-17] The existence of zero modes and the Landau level structure have been established also for tilted Dirac cones. However, the effect of disorder, in particular the robustness of zero modes and the role of symmetry, has not been explored until recently.[18] In the present paper, with an explicit form of the eigenstates of the generalized chiral operator, we demonstrate how the chiral symmetry, which is broken for tilted cones, can be generalized to include tilted Dirac cones and clarify its relevance to the robustness of the zero modes of generic massless Dirac fermions. Relationships between the generalized chiral symmetry and the applicability of the index theorem is also elaborated.

2. Generalization of Chiral Symmetry

To illustrate how the chiral symmetry is generalized to tilted Dirac cones, let us consider a general form of the Hamiltonian,[18]

$$H(\eta) = -\eta v_F \sigma_0 \pi_x + v_F(\sigma_x \pi_x + \sigma_y \pi_y),$$

for a two-dimensional massless Dirac fermion in a magnetic field, where the isotropic Dirac cone is tilted in the x direction for $\eta \neq 0$ with σ_0 being the unit matrix. The dynamical momentum $\boldsymbol{\pi}$ satisfies the commutation relation $[\pi_x, \pi_y] = -ie\hbar B$ with $\boldsymbol{B} = \text{rot}\boldsymbol{A}$. The parameter η determines the tilting of the Dirac cone. Note that for $\boldsymbol{A} = 0$ an equienergy contour of the Dirac cone is elliptic as long as $|\eta| < 1$ (while hyperbolic for $|\eta| > 1$; see Fig.1(a)). The first term in the Hamiltonian destroys the chiral symmetry as $\sigma_z H(\eta) \sigma_z = -H(-\eta) \neq -H(\eta)$. We can, however, define a generalized chiral operator γ as

$$\gamma = \lambda_\eta^{-1}(\sigma_z - i\eta\sigma_y), \quad \lambda_\eta = \sqrt{1-\eta^2},$$

which arises naturally in the general framework of the eigenvalue problem for tilted Dirac cones.[19] Although the generalized chiral operator γ is not hermitian ($\gamma^\dagger \neq \gamma$), one can readily verify that $\gamma^2 = 1$ with eigenvalues ± 1. The corresponding right-eigenvectors $\gamma|\pm\rangle = \pm|\pm\rangle$ are given explicitly as

$$|+\rangle = \frac{1}{\sqrt{2(1+\lambda_\eta)}}\begin{pmatrix} 1+\lambda_\eta \\ \eta \end{pmatrix}, \quad |-\rangle = \frac{1}{\sqrt{2(1+\lambda_\eta)}}\begin{pmatrix} \eta \\ 1+\lambda_\eta \end{pmatrix}.$$

In the limit of the vertical cone ($\eta \to 0$), the generalized operator γ reduces to the conventional chiral operator σ_z. With the generalized chiral operator, we find that for $|\eta| < 1$,

$$\gamma^\dagger H(\eta)\gamma = -H(\eta),$$

which we call the *generalized chiral symmetry*.[18] The generalized symmetry guarantees the identity $\langle +|H(\eta)|+\rangle = \langle -|H(\eta)|-\rangle = 0$, and plays an essential role for the robustness of zero modes as shown in the next section. Note that this symmetry holds irrespective of the details of the vector potential $\mathbf{A}$. Disorder in gauge degrees of freedom (such as random magnetic fields) respects this symmetry.

3. Robustness of Zero Modes

If the above generalized chiral symmetry is preserved, the Schrödinger equation $H(\eta)\psi = E\psi$ for the wave function $\psi = |+\rangle\psi^+ + |-\rangle\psi^-$ becomes

$$\begin{pmatrix} 0 & \lambda_\eta \pi_x - i\pi_y \\ \lambda_\eta \pi_x + i\pi_y & 0 \end{pmatrix}\begin{pmatrix} \psi^+ \\ \psi^- \end{pmatrix} = E\begin{pmatrix} 1/\lambda_\eta & \eta/\lambda_\eta \\ \eta/\lambda_\eta & 1/\lambda_\eta \end{pmatrix}\begin{pmatrix} \psi^+ \\ \psi^- \end{pmatrix}.$$

The zero ($E = 0$) modes are then given by the wave functions satisfying[18]

$$(\lambda_\eta \pi_x - i\pi_y)\psi^- = 0, \ \psi^+ = 0 \qquad (1)$$

or

$$(\lambda_\eta \pi_x + i\pi_y)\psi^+ = 0, \ \psi^- = 0. \qquad (2)$$

The zero modes are thus the eigenstates of the generalized chiral operator γ and have either "−" chirality with $\gamma\psi = -\psi$ (Eq.(1)) or "+" chirality with $\gamma\psi = \psi$ (Eq.(2)). It is to be recalled that for a vertical Dirac cone the zero modes are also the eigenstates of the chiral operator $\Gamma = \sigma_z$. These equations for the zero modes hold even in the case where the gauge field is disordered.

Following Aharonov and Casher,[10] we adopt the "Coulomb gauge" $\lambda_\eta \partial_x A_x + \lambda_\eta^{-1}\partial_y A_y = 0$ by assuming $\mathbf{A} = (-\lambda_\eta^{-1}\partial_y\varphi, \lambda_\eta \partial_x\varphi)$. Then Eq.(1) for ψ^- and Eq.(2) for ψ^+ are reduced to

$$[D_\pm \mp (e/\hbar)(D_\pm\varphi)]\psi^\pm = 0,$$

where $D_\pm \equiv (\partial_X \pm i\partial_Y)$ with $\mathbf{R} = (X, Y) = (x/\sqrt{\lambda_\eta}, y\sqrt{\lambda_\eta})$. The solutios are then given by $\psi^\pm = \exp(\pm e\varphi/\hbar)f(Z_\pm)$ with a polynomial $f(Z_\pm)$ in $Z_\pm \equiv X \pm iY$. Since $(\partial_X^2 + \partial_Y^2)\varphi = B$, we have $\varphi(\mathbf{R}) = \int d\mathbf{R}' G(\mathbf{R} - \mathbf{R}')B(\mathbf{R}')$ where $G(\mathbf{R}) = (1/2\pi)\log(R)$ with $R = \sqrt{X^2 + Y^2}$, which leads to the asymptotic form $\varphi \to (\Phi/2\pi)\log R$ in the limit as $R \to \infty$ with Φ being the total magnetic flux in the system. We then have to chose the chirality for the wave function to be normalizable. For instance, when the total magnetic flux Φ is positive, the zero mode has to have "$-$" chirality to be normalizable. The number of square-normalizable wave function then becomes $\Phi/(h/e)$,[10] that exactly coincides with the degeneracy of the Landau level. This implies the zero-modes exhaust the Landau level of a tilted Dirac cone so that its density of states is a *generalized chiral symmetry protected delta-function* in the presence of disorder. In this sense, the generalized chiral symmetry protects the zero mode of a generic massless Dirac fermions in two dimensions.

4. Generalized Chiral Symmetry and Index Theorem

The above reasoning gives an explicit relationship between the index theorem and the generalized chiral symmetry. The index theorem, which gives the least number of zero modes, holds for an elliptic operator.[8] In the present case, the Hamiltonian $H(\eta)$ is elliptic as a differential operator if the matrix,

$$\Xi(\xi_x, \xi_y) = \begin{pmatrix} -\eta\xi_x & \xi_x - i\xi_y \\ \xi_x + i\xi_y & -\eta\xi_x \end{pmatrix},$$

is invertible for any $(\xi_x, \xi_y) \in \mathbf{R}^2 - (0,0)$. The condition for the ellipticity thus becomes $\det\Xi = -(1-\eta^2)\xi_x^2 - \xi_y^2 \neq 0$. This determinant is always negative ($\det\Xi < 0$) and never becomes zero for $(\xi_x, \xi_y) \in \mathbf{R}^2 - (0,0)$, as long as $|\eta| < 1$. The condition $|\eta| < 1$ is therefore exactly the condition for the ellipticity of the Hamiltonian $H(\eta)$ as well as that for the existence of the generalized chiral symmetry. In this sense, *the generalized chiral symmetry is generically equivalent to the ellipticity of the Dirac-cone Hamiltonian* with the same parameter space for their validity. Geometrically, the condition $|\eta| < 1$ means that the Dirac cone is not tilted too much so that the cross section with a constant energy plane remains to be elliptic.

5. Numerical Demonstration

We have also preformed numerical calculations based on a lattice model having a pair of tilted Dirac cones at $E = 0$ as shown in Fig.1(b)(Inset). The model is defined on the two-dimensional square lattice with a Hamiltonian having the nearest-neighbor (t) and the next-nearest neighbor (t') transfer integrals as[18,20]

$$H = \sum_{\bm{r}} -tc^\dagger_{\bm{r}+\bm{e}_x}c_{\bm{r}} + (-1)^{x+y}tc^\dagger_{\bm{r}+\bm{e}_y}c_{\bm{r}} + t'(c^\dagger_{\bm{r}+\bm{e}_x+\bm{e}_y}c_{\bm{r}} + c^\dagger_{\bm{r}+\bm{e}_x-\bm{e}_y}c_{\bm{r}}) + \text{H.c.},$$

where $\bm{r} = (x,y)$ denotes the lattice point in units of the nearest-neighbor distance a, and $\bm{e}_x(\bm{e}_y)$ the unit vector in the $x(y)$ direction. The magnetic field is taken into account by the Peierls phases $\theta(\bm{r})$ as $t(t') \to t(t')\exp(-2\pi i\theta(\bm{r}))$, so that the their sum along the closed loop is equal to the enclosed magnetic flux in units of the flux quantum h/e. To see how the Landau levels are broadened by disorder in the present model, we assume that the magnetic flux $\phi(\bm{r})$ enclosed by the square located at $\bm{r}$ has a random component $\delta\phi(\bm{r})$ in addition to the uniform component

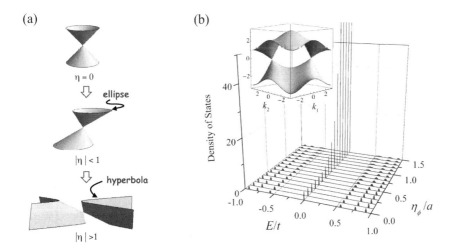

Fig. 1. (a) Schematic figures of the Dirac cones with $\eta = 0$, $|\eta| < 1$ and $|\eta| > 1$. (b) Density of states for the lattice model in a magnetic field with various values of the correlation length η_ϕ of the random component of the magnetic flux. Here, the parameters are assumed to be $t'/t = 0.25$, $\phi = 0.01(h/e)$, and $\sqrt{\langle\delta\phi^2\rangle} = 0.0029(h/e)$. The system-size is $20a$ by $20a$ and the average over 10^4 samples is made. Inset: Energy dispersion $E(\mathbf{k})/t$ in the absence of a magnetic field, where $k_1 = \mathbf{k} \cdot (\mathbf{e}_x + \mathbf{e}_y)$ and $k_2 = \mathbf{k} \cdot (\mathbf{e}_x - \mathbf{e}_y)$.

ϕ. The random components have a gaussian distribution and are correlated in space as $\langle\delta\phi(\mathbf{r}_1)\delta\phi(\mathbf{r}_2)\rangle = \langle\delta\phi^2\rangle\exp(-|\mathbf{r}_1 - \mathbf{r}_2|^2/4\eta_\phi^2)$. The disorder in magnetic fluxes should appear, for large η_ϕ, as the disorder in gauge degrees of freedom, and therefore should respect the generalized chiral symmetry of the effective Hamiltonian at the Dirac points. As shown in Fig.1(b), we actually see an anomalously sharp $n = 0$ Landau level when the correlation length η_ϕ of disorder becomes larger than the nearest-neighbor distance a, while other Landau levels ($n = \pm 1, \pm 2, \pm 3$) are broadened as usual. Together with the result[18] for a different inclination of the cone, this anomaly at the $n = 0$ Landau level is likely to exist universally, irrespective of the magnitude of the tilting angle of the cone. This anomalously sharp $n = 0$ Landau level suggests that the energy levels are exactly degenerated at $E = 0$. The present numerical results are thus consistent with the robustness of zero modes of a single tilted Dirac cone protected by the generalized chiral symmetry.

6. Conclusions

We have shown explicitly that the notion of the chiral symmetry can be generalized to generic tilted Dirac cones, where the generalized chiral operator has to be non-hermitian. It has been also demonstrated analytically that the generalized chiral symmetry indeed protects the zero modes of the system by extending the argument by Aharonov and Casher.[10] The resulting anomalously sharp $n = 0$ Landau level has been also confirmed numerically based on the lattice model. The existence of the generalized chiral symmetry coincides with the ellipticity of the Hamiltonian as a differential operator, which is nothing but the geometrical condition that the Dirac cone is not tilted too much so that the cross section with a constant energy plane is an ellipse.

Acknowledgments

The authors wish to thank Yoshiyuki Ono and Tomi Ohtsuki for useful discussions. This work was partly supported by Grants-in-Aid for Scientific Research, Nos. 22540336 and 23340112 from JSPS.

References

1. A.W.W. Ludwig, et al., *Phys. Rev. B* **50**, 7526 (1994).
2. A. Altland and M.R. Zirnbauer, *Phys.Rev. B* **55**, 1142 (1997).
3. F. Evers and A.D. Mirlin, *Rev. Mod. Phys.* **80**, 1355 (2008).
4. K.S. Novoselov et al, *Nature* **438**, 197 (2005).
5. Y. Zhang et al., *Nature* **438**, 201 (2005).
6. A.H. Castro Neto et al, *Rev. Mod. Phys.* **81**, 109 (2009).
7. P.M. Ostrovsky, I.V. Gornyi, and A.D. Mirlin, *Phys. Rev. B* **77**, 195430 (2008).
8. M. Nakahara, *Geometry, Topology, and Physics*, 2nd ed. (Taylor & Francis, 2003).
9. M.I. Katsnelson and K.S. Novoselov, *Solid State Commun.* **143**, 3 (2007).
10. Y. Aharonov and A. Casher, *Phys. Rev. A* **19**, 2461 (1979).
11. J. Kailasvuori, *Europhys. Lett.* **87**, 47008 (2009).
12. T. Kawarabayashi, Y. Hatsugai, and H. Aoki, *Phys. Rev. Lett.* **103**, 156804 (2009); *Physica E***42**, 759 (2010); T. Kawarabayashi, T. Morimoto, Y. Hatsugai, and H. Aoki, *Phys. Rev. B* **82**, 195426 (2010).
13. S. Katayama, A. Kobayashi, and Y. Suzuura, *J. Phys. Soc. Jpn.* **75**, 054705 (2006).
14. N. Tajima, S. Sugawara, M. Tamura, Y. Nishio, and K. Kajita, *J. Phys. Soc. Jpn.* **75**, 051010 (2006).
15. T. Morinari, T. Himura and T. Tohyama, *J. Phys. Soc. Jpn.* **78**, 023704 (2009).
16. T. Morinari and T. Tohyama, *J. Phys. Soc. Jpn.* **79**, 044708 (2010).
17. M.O. Goerbig, J.-N. Fuchs, G. Montambaux, and F. Piéchon, *Phys. Rev. B* **78**, 045415 (2008).
18. T. Kawarabayashi, Y. Hatsugai, T. Morimoto, and H. Aoki, *Phys. Rev. B* **83**, 153414 (2011).
19. Y. Hatsugai, T. Kawarabayashi, and H. Aoki, in preparation.
20. Y. Morita and Y. Hatsugai, *Phys. Rev. Lett.* **79**, 3728 (1997); Y. Hatsugai, X.-G. Wen, and M. Kohmoto, *Phys. Rev. B* **56**, 1061 (1997).

ELECTRONIC STATES AND LOCAL DENSITY OF STATES NEAR GRAPHENE CORNER EDGE

YUJI SHIMOMURA

Department of Quantum Matter, AdSM, Hiroshima University,
Higashi-Hiroshima, Hiroshima, 739-8530, Japan
shimomura25@hiroshima-u.ac.jp

YOSITAKE TAKANE

Department of Quantum Matter, AdSM, Hiroshima University,
Higashi-Hiroshima, Hiroshima, 739-8530, Japan

KATSUNORI WAKABAYASHI

International Center for Materials Nanoarchitectonics (WPI-MANA),
National Institute for Materials Science (NIMS),
Namiki 1-1, Tsukuba 305-0044, Japan

We study that stability of edge localized states in semi-infinite graphene with a corner edge of the angles 60°, 90°, 120° and 150°. We adopt a nearest-neighbor tight-binding model to calculate the local density of states (LDOS) near each corner edge using Haydock's recursion method. The results of the LDOS indicate that the edge localized states stably exist near the 60°, 90°, and 150° corner, but locally disappear near the 120° corner. By constructing wave functions for a graphene ribbon with three 120° corners, we show that the local disappearance of the LDOS is caused by destructive interference of edge states and evanescent waves.

Keywords: Graphene corner edge; zigzag; edge localized.

PACS numbers: 81.05.ue, 73.22.Pr, 73.20.-r

1. Introduction

Graphene is the first true two-dimensional material composed of carbon atoms.[1] Since it contains non-equivalent two carbon atoms in the unit cell due to the nature of honeycomb lattice, the low-energy electronic properties of graphene are well described by the massless Dirac equation, where the valence and conduction bands conically touch at the Fermi energy.

The presence of edges in graphene makes strong implications for electronic states near the Fermi energy. There are two representative edges structures in graphene, called zigzag (zz) and armchair (ac). Zigzag edge provides highly degenerated edge localized state (edge state) at the Fermi energy,[2] but armchair edge does not show such localized state. Recently the atomic structures of graphene edges were studied using scanning tunneling microscope and successively the presence of edge states is confirmed using scanning tunneling spectroscopy.[3,4] Although in these experiments the relatively long armchair edges are observed, zigzag edges preferably appear to form corner edges.[3]

In our previous publication,[5] we have shown that electronic states of graphene with a corner edge crucially depend on the corner angle. Corners with the angles 60°, 90° and 150° can stably possess the edge states, where the wave functions can be analytically obtained as a simple linear combination of edge states.[5,6] In the 120° case, however, the edge states locally disappear, and an appropriate wave function cannot be obtained in a similar manner. In this paper, we will numerically demonstrate that the wave functions of 120° corner edge can be described as the destructive interference between the edge states and evanescent waves.

2. LDOS on Corner Edges

We assume that π electronic states in graphene are described by a tight-binding model on a honeycomb lattice. The Hamiltonian is $H = -t \sum_{\langle i,j \rangle} |i\rangle\langle j|$, where $\langle i,j \rangle$ is a pair of the nearest-neighbor sites, and t is a hopping integral. For a given energy E, we calculate the LDOS at an arbitrary site using Haydock's recursion method.[5,7–10] Throughout this paper, the origin of energy is set to the Fermi energy of graphene. To calculate the LDOS at the ith size, we transform our tight-binding model to a one-dimensional chain model with a tridiagonal Hamiltonian[a]:

$$\mathcal{H} = \begin{pmatrix} a_0 & b_1 & & \\ b_1 & a_1 & b_2 & \\ & b_2 & a_2 & \\ & & & \ddots \end{pmatrix}. \quad (1)$$

The Green's function $G_i(E)$ for the ith site is approximately expressed as

$$G_i(E) = \cfrac{1}{E - a_0 - \cfrac{b_1^2}{E - a_1 - \cfrac{b_2^2}{E - a_N - t(E)}}}, \quad (2)$$

where

$$t(E) = \frac{E - a_N}{2b_N^2} \left[1 - \left\{ 1 - \frac{4b_N^2}{(E - a_N)^2} \right\}^{\frac{1}{2}} \right]. \quad (3)$$

Eq. (2) exactly takes account of the effects of all carbon sites up to the Nth neighbor sites from ith site. The effects of other carbon sites beyond this distance are

---

[a]The normalized bases of $\mathcal{H}$ are determined in terms of the recursion relation $b_{n+1}|n+1\rangle = (H - a_n)|n\rangle - b_n|n-1\rangle$ with $a_n = \langle n|H|n \rangle$ and the initial state $|0\rangle \equiv |i\rangle$. See, Ref. 5 for details.

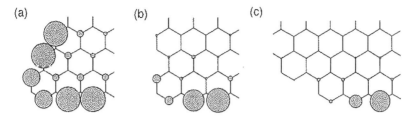

Fig. 1. The LDOS at $E = 0$ of semi-infinite graphene with a corner edge with angles of (a) 60°, (b) 90°, and (c) 150°. The radius of circles denotes the magnitudes of LDOS.

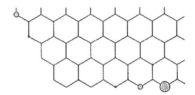

Fig. 2. The LDOS at $E = 0$ for semi-infinite graphene with a corner edge of the 120°.

approximately taken into account through $t(E)$. $G_i(E)$ gives the LDOS at the ith site in terms of the relation

$$n_i(E) = \frac{1}{\pi}\mathrm{Im}G_i(E - i\delta), \qquad (4)$$

where δ is a positive infinitesimal. We calculate the LDOS in the case of 60°, 90°, 120° and 150° corner edges by setting $N = 1000$.

Figure 1 shows the LDOS for the corner edges with 60°, 90° and 150°, while Fig. 2 shows the LDOS for the corner edge with 120°. Here the energy is fixed at the Fermi energy of graphene, i.e. $E = 0$. In the cases of 60°, 90° and 150°, we observe the characteristic feature of the edge states that the LDOS have the finite amplitudes only on one of two sub-lattices. In contrast, in the case of 120°, the LDOS at zero energy almost disappears near the corner edge, but its value on the zz edge increases with being distant from the corner.

3. Wave Functions for 120° Corner

It is interesting to consider wave functions that can account for the behavior of the LDOS observed in the previous section. In the 60°, 90° and 150° cases, such wave functions can be analytically obtained as a simple linear combination of edge states.[5,6] In the 120° case, however, an appropriate wave function cannot be obtained in a similar manner. This implies that not only the edge modes but also locally induced evanescent waves play a role.

To numerically construct wave functions for 120° corner, we propose to treat a semi-infinite zz ribbon having three 120° corner edges as shown in Fig. 3(a). Since we are interested in the local disappearance of edge states near 120° corner, our

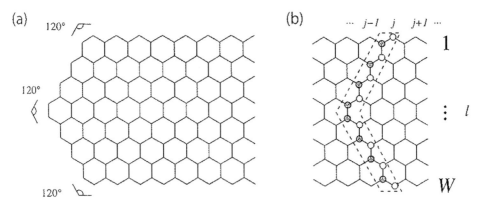

Fig. 3. (a) Schematic structure of semi-infinite zz graphene naoribbon, where left-side is terminated by three corner edges of 120° degrees. (b) The polygon with dashed lines defines the unit cell. Since each zz line in the unit cell contains two carbon sites, we refer to the left (right) site on the lth zz line as $l\alpha(l\beta)$. The black (white) circles denote $\alpha(\beta)$ sites.

attention is focused on low-energy states with $E \cong 0$. We take the unit cell of the zz ribbon as shown in Fig. 3(b). Each unit cell has $2W$ sites, where W is the number of zz lines. The jth unit cell contains two sites of each zz line. We refer to the left (right) site on the lth zz line as $l\alpha(l\beta)$. We define $\mathbf{C}_j^\alpha(\mathbf{C}_j^\beta)$ as the column vector consisting of the wave coefficients at $l\alpha(l\beta)(l = 1, 2, \ldots, W)$. The column vectors $\mathbf{C}_j^\alpha$ and $\mathbf{C}_j^\beta$ satisfy the relations

$$\mathbf{H}_1 \mathbf{C}_j^\alpha + \mathbf{H}_2 \mathbf{C}_j^\beta + \mathbf{H}_2 \mathbf{C}_{j-1}^\beta = E \mathbf{C}_j^\alpha, \tag{5}$$

$$\mathbf{H}_2^\dagger \mathbf{C}_j^\alpha + \mathbf{H}_3 \mathbf{C}_{j+1}^\alpha = E \mathbf{C}_j^\beta, \tag{6}$$

where $\mathbf{H}_1$ ($\mathbf{H}_2$) represents the intra-cell transfer between nearest neighbor α (α and β) sites in jth unit cell, and $\mathbf{H}_3$ represents the inter-cell transfer between nearest neighbor α and β sites, respectively. Eliminating terms with the column vectors for β sites from (5) and (6), we obtain

$$\mathbf{C}_{j+1}^\alpha + \mathbf{u}\mathbf{C}_{j-1}^\alpha + \mathbf{v}\mathbf{C}_j^\alpha = 0, \tag{7}$$

where $\mathbf{u}$ and $\mathbf{v}$ are matrices which are constructed in terms of $\mathbf{H}_1$, $\mathbf{H}_2$, $\mathbf{H}_3$ and E. Here, we assume that the column vector satisfies $\mathbf{C}_{j+1}^\alpha = \lambda \mathbf{C}_j^\alpha$. From (7), the eigenvalue equation

$$\begin{pmatrix} -\mathbf{v} & -\mathbf{u} \\ 1 & 0 \end{pmatrix} \begin{pmatrix} \mathbf{C}_j^\alpha \\ \mathbf{C}_{j-1}^\alpha \end{pmatrix} = \lambda \begin{pmatrix} \mathbf{C}_j^\alpha \\ \mathbf{C}_{j-1}^\alpha \end{pmatrix} \tag{8}$$

is derived. Solving (8), we obtain $2W$ eigenmodes. They can be classified into three types according to the absolute value of an eigenvalue λ. The first type is the eigenmodes with $|\lambda| = 1$. At $E \cong 0$, we obtain two such eigenmodes representing edge states. One is the right-going mode, and the other is the left-going mode. The second type is the evanescent modes with $|\lambda| < 1$, and the number of these modes is $W - 1$. The last type is the modes with $|\lambda| > 1$ which exponentially increase toward the right direction. The number of these eigenmodes is also $W-1$. We have assumed that the zz ribbon is infinitely long to the right direction (see Fig. 3), so the eigenmodes of the last type should be excluded because they diverges with increasing j. Thus we take the two edge modes of the first type and the $W-1$ evanescent modes of the second type as the basis functions.

Let us consider a scattering problem in the case where the left-going edge mode is incident from the right. Then, the right-going edge mode is regarded as a reflected wave, and $W-1$ evanescent modes play a role of scattered waves. Let us express the two edge modes as $\psi_{\text{in}}(\mathbf{r})$ and $\psi_{\text{ref}}(\mathbf{r})$, and the $W-1$ evanescent modes as $\psi_1(\mathbf{r}), \ldots, \psi_{W-1}(\mathbf{r})$, where $\mathbf{r}$ represents the site specified by j, l and η with $\eta = \alpha$ or β. In the zz ribbon system, the wave function $\Psi(\mathbf{r})$ is represented as a linear combination of these waves

$$\Psi(\mathbf{r}) = \psi_{\text{in}}(\mathbf{r}) + a\psi_{\text{ref}}(\mathbf{r}) + \sum_{k=1}^{W-1} b_k \psi_k(\mathbf{r}), \qquad (9)$$

where a and $\{b_k\}$ are unknown coefficients to be determined. We consider the boundary condition at the left boundary consisting of three $120°$ corner edges. Note that this boundary can be created from an infinitely long zz ribbon by removing W sites, each of which is the nearest neighbor of a boundary site. We require that the wave function vanishes at these removed sites, i.e., $\Psi(\mathbf{r}_1) = \Psi(\mathbf{r}_2) = \cdots = \Psi(\mathbf{r}_W) = 0$, where $\mathbf{r}_1, \ldots, \mathbf{r}_W$ are positions of the removed sites. Imposing this boundary condition to (9), we numerically determine the unknown coefficients. Actual calculation is performed for the case of $W = 30$.

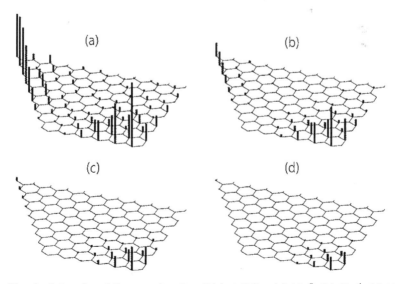

Fig. 4. The absolute value of the wave function $\Psi(\mathbf{r})$ at $E/l =$ (a) 10^{-3}, (b) 10^{-4}, (c) 10^{-5}, and (d) 10^{-6} near one of the $120°$ corner edges. The length of bars indicates $|\Psi(\mathbf{r})|$.

In Fig. 4, we show the absolute value of the wave function $|\Psi(\mathbf{r})|$ at $E/t =$ (a) 10^{-3}, (b) 10^{-4}, (c) 10^{-5} and (d) 10^{-6} near the 120° corner edge. The figures show an extracted area from the entire zz ribbon. We observe that $|\psi(\mathbf{r})|$ almost disappears near the corner, but its amplitude on the zz edge increases with being distant from the corner edge. These features are consistent with the behavior of LDOS.

4. Summary

We have studied electronic states of semi-infinite graphene with a corner edge, focusing on the stability of edge localized states. We have calculated the local density of states (LDOS) of each case based on the nearest-neighbor tight-binding model using Haydock's recursion method. The edge localized states stably exist near the 60°, 90° and 150° corners, but locally disappear near the 120° corner. By constructing wave functions for a graphene ribbon with three 120° corners, we found that the local disappearance of the LDOS is caused by destructive interference of edge states and evanescent waves. Recent experiments has reported that zigzag edges with many ~120° turns had been observed on the graphene grains synthesized on Cu foils by chemical vapor deposition.[11] The local disappearance of edge localized states for 120° corner edge might be related to the above experiment, because the local disappearance of edge localized states energetically stabilizes the formation of corner edges with such angle. Further theoretical details will be published elsewhere.

References

1. A. K. Geim and K. S. Novoselov, *Nature Mat.* **6**, 183 (2007).
2. M. Fujita, K. Wakabayashi, K. Nakada and K. Kusakabe, *J. Phys. Soc. Jpn.* **65**, 1920 (1996).
3. Y. Kobayashi, K. Fukui, T. Enoki, K. Kusakabe and Y. Kaburagi, *Phys. Rev. B* **71**, 193406 (2005).
4. Y. Niimi, T. Matsui, H. Kambara, K. Tagami, M. Tsukada and H. Fukuyama, *Phys. Rev. B* **73**, 085421 (2006).
5. Y. Shimomura, Y. Takane and K. Wakabayashi, *J. Phys. Soc. Jpn.* **80**, 054710 (2011).
6. M. Ezawa, *Phys. Rev. B* **81**, 201402(R) (2010).
7. R. Haydock, The recursive solution of the Schrödinger equation, in *Solid State Physics*, Vol. 35 (Academic, New York, 1980), p. 216.
8. M. J. Kelly, The recursive solution of the Schrödinger equation, in *Solid State Physics*, Vol. 35 (Academic, New York, 1980), p. 296.
9. L. C. Davis, *Phys. Rev. B* **28**, 6961 (1983).
10. S. Wu, L. Jing, Q. Li, Q. W. Shi, J. Chen, H. Su, X. Wang and J. Yang, *Phys. Rev. B* **77**, 195411 (2008).
11. J. Tian, H. Cao, W. Wu and Y. P. Chen, *Nano Lett.* **11**, 3363 (2011).

PERFECTLY CONDUCTING CHANNEL AND ITS ROBUSTNESS IN DISORDERED CARBON NANOSTRUCTURES

YUKI ASHITANI[1], KEN-ICHIRO IMURA[1,2], YOSITAKE TAKANE[1]*

[1]*Department of Quantum Matter, Graduate School of Advanced Sciences of Matter, Hiroshima University, Higashihiroshima, Hiroshima, 739-8530, Japan;* [2]*Kavli Institute for Theoretical Physics, University of California, Santa Barbara, CA 93106, USA.*

We report our recent numerical study on the effects of dephasing on a perfectly conducting channel (PCC), its presence believed to be dominant in the transport characteristics of a zigzag graphene nanoribbons (GNR) and of a metallic carbon nanotubes (CNT). Our data confirms an earlier prediction that a PCC in GNR exhibits a peculiar robustness against dephasing, in contrast to that of the CNT. By studying the behavior of the conductance as a function of the system's length we show that dephasing destroys the PCC in CNT, whereas it stabilizes the PCC in GNR. Such opposing responses of the PCC against dephasing stem from a different nature of the PCC in these systems.

Keywords: electron transport; carbon nanotube; graphene nanoribbon; dephasing.

PACS numbers: 72.80.Vp, 73.20.Fz

1. Introduction

In one-spatial dimension any weak disorder is believed to have the potentiality of converting a good metal to an insulator, as a consequence of the Anderson localization.[1] However, the existence of two counter examples for this common belief has been pointed out recently, both being found in a carbon nanostructure, exhibiting a perfectly conducting channel (PCC). The two examples are i) the metallic carbon nanotube (CNT),[2–5] and ii) the zigzag graphene nanoribbon (GNR) with edge modes of partially flat dispersion.[6–8] A PCC is immune to backward scattering; its existence allowing the conductance of the system to remain *finite* even when its length L becomes infinitely long, indicating the *absence* of Anderson localization. Note also that both CNT and GNR can be regarded as a derivative form of an infinitely large graphene sheet possessing two energy valleys around its Dirac points K and K'. Since scattering between these two valleys, i.e., the inter-valley scattering, usually destroys the perfectly conducting channel, we focus on the case in which the system is subject to only long-ranged scatterers, i.e., impurities whose potential range is larger than or comparable to the size of the unit cell.

---

*Author to whom correspondence should be addressed.

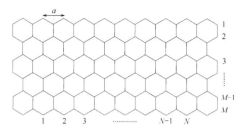

Fig. 1. Real space image of a GNR consisting of M zigzag lines. A CNT consisting of the same number of zigzag lines can be obtained by rolling up this GNR and linking each site of the first zigzag line with its partner on the Mth row.

This paper highlights the behavior of such a PCC believed to be existent in the carbon nanostructures. Since a PCC appears within a quantum-mechanical framework, one may think that it is fragile against a loss of the phase coherence due to inter-electronic Coulombic interaction, electron-phonon coupling, etc. This naive speculation, however, turns out to be not necessarily the case, as we further elaborate the description of this phenomenon below. We have performed extensive numerical study of such carbon-based disordered quasi-one-dimensional systems using the standard tight-binding representation of the graphene's honeycomb lattice structure (see Fig. 1). Our treatment of the dephasing follows that of Ref. 5.

2. Perfectly conducting channels in GNR and CNT

In the case of GNR with zigzag edge boundaries, the existence of a PCC is originated from its peculiar band structure. Indeed, one can give it a simple interpretation based on the appearance of partially flat-band edge modes.[9] Since these flat bands appear only in a part of the one-dimensional Brillouin zone connecting the two valleys, if one counts the number of conducting channels of each propagating direction at a given Fermi energy, there always exist an excess right-going channel in one valley and an excess left-going channel in the other valley. Let N_c be the number of conducting channels in each valley in the absence of the edge modes. The above fact indicates that the number of right-going (left-going) channels is N_c ($N_c + 1$) in one valley and $N_c + 1$ (N_c) in the other valley. This imbalance leads to the appearance of one PCC which is robust against disorder,[10,11] resulting in a noteworthy statement on the scaling of the dimensionless conductance $g(L)$, i.e., "$g(L)$ scales naturally to a smaller value as the length L of the disordered region increases, but in the large-L limit, $g(L)$ remains to be a finite as $\lim_{L \to \infty} g(L) = 1$.[6–8] Interested readers may refer to Ref. 12 and references therein for more detailed discussion on the transport characteristics of such a system with an imbalance in the number of right- and left-going channels. In a recent paper,[12] one of the authors has shown that this PCC still survives even in the incoherent regime, where information on the phase of the electronic wave function is essentially lost. This unexpected robustness of the PCC in GNRs in the presence of dephasing (see also Fig. 2) stems most certainly from the

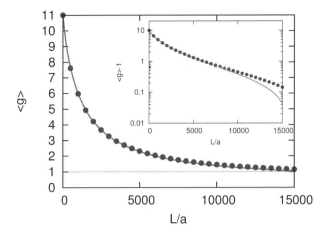

Fig. 2. Conductance of a disordered graphene nanoribbon: a linear plot of the dimensionless conductance $\langle g \rangle$ (main panel), and a semi-log plot of $\langle g \rangle - 1$ (inset) as a function of the length L of the disordered region measured in units of the lattice constant a. Solid lines (filled dots) corresponds to the case without (with) dephasing. We set $M = 30$ and $\epsilon_F/t = 0.579$ for which the total number of conducting channels is 11 (i.e., $g = 11$ at $L/a \to 0$). Other parameters are $W/t = 0.13$, $p = 0.1$ and $L_\phi/a = 500$, where W measures the strength of each scatterer, and p is the probability that each site is occupied by a such scatterer. The ensemble average is performed over 10^4 samples with different impurity configurations. The magnitude of the error bar at $L/a = 15000$ is of order 10^{-3}.

fact that the imbalance in the number of conducting channels is not a consequence of a particular symmetry (cf. role of the so-called "pseudo-time reversal" symmetry in the CNT case, see the discussion below); it is simply guaranteed by the existence of partially flat-band edge modes.

In contrast to the case of GNR, the existence of a PCC in CNTs is a much subtle issue. It is certainly essential that the system belongs to the symplectic symmetry class, i.e., the total Hamiltonian of the system inclusive of the random potential must, not only be time-reversal symmetric (TRS), but also fall on the case of $\Theta^2 = -1$ with Θ being the time-reversal operator. This is typically the case with an effective spin-1/2 system of a Dirac-type conic dispersion relation though in this case TRS is not a real one (often dubbed as "pseudo-TRS"). This condition, therefore, will be safely satisfied in CNTs under the influence of long-ranged potential disorder. However, this condition alone turns out to be still not a sufficient one for ensuring the existence of a PCC. Much work on this subtlety, associated with the parity of the number N_c of the conducting channels in each single Dirac cone, has been pursued by Ando and co-workers in the context of studying the transport characteristics of CNTs at a very early stage in the development of this field.[2-4] To the best of our knowledge a clear statement on the condition for the appearance of PCC, i.e., the idea that *both* of the following two conditions: i) appurtenance to the symplectic symmetric class, ii) oddness of the number of conducting channels, must be satisfied, has first appeared in Ref. 5. Notice that in the band structure of metallic CNTs, only the single lowest gapless subband (of quasi-linear dispersion) is non-degenerate, whereas other quadratic subbands are all two-fold degenerate. Therefore, wherever the Fermi level ϵ_F is, the number of conducting channels in each propagating direction in a given valley is necessarily odd. This ensures the existence of at least one PCC per valley (cf. Fig. 3). Clearly, the dimensionless conductance g ($= 2N_c$ in the clean limit; here the factor 2 comes from the two valleys) decreases as disorder increases, but remains finite due to the appearance of two PCCs. This can be rephrased as follows: "For a fixed strength of disorder, g scales down to a smaller value as the system becomes longer (as L increases), but it approaches asymptotically to an integral value, which is 2, in the long-L limit". Such a behavior of the so-called "symplectic-odd symmetry class" has been more profoundly elucidated by the subsequent studies[13–15] in the context of the DMPK equation and the supersymmetric field theory.[a] The existence of PCC in CNTs relies on the presence of pseudo-TRS. Therefore, it could be fragile against any disturbances that might cause breaking of the pseudo-TRS, e.g., against trigonal warping of the Dirac cone.[18] It is, therefore, natural to presume that PCC might be fragile against dephasing.[5] In this paper, we have extended this consideration on the role of dephasing in the robustness of PCC in CNTs, primarily for the comparison with the GNR case, but with much care to the dependence on the circumference R of the nanotube.[b]

3. Sketch of the numerical analysis and its implications

Let us consider again the case of a GNR with M zigzag lines as shown in Fig.1. The electronic states in this nanostructure is described by a tight-binding Hamiltonian,

$$H = -\sum_{i,j} \gamma_{i,j} |i\rangle\langle j| + \sum_i V_i |i\rangle\langle i|, \tag{1}$$

where $|i\rangle$ and V_i represent the localized electron state and the impurity potential, respectively, on site i, and $\gamma_{i,j}$ is the transfer integral between sites i and j with $\gamma_{i,j} = t$ if i and j are nearest neighbors and $\gamma_{i,j} = 0$ otherwise. We assume that the zigzag lines are infinitely long. Instead, we distribute impurities (randomly) only in a finite region (the disordered region, composed of N columns) of this infinitely long ribbon. What we have been calling the "system's length L" so far is now identified as the length N of this disordered region, i.e., $L/a = N$.

In the actual computation, we have numerically estimated the dimensionless conductance $g(L)$ using the Landauer formula and recursive Green's function method.

---

[a]It seems fair to mention that a similar idea but in a different context has already appeared in a earlier work of Zirnbauer and co-workers.[16,17]

[b]The larger the circumference R is, the more closely are the subbands spaced. Also, the further one goes away from the Dirac point, the stronger the trigonal warping becomes in the spectrum of a CNT. Combining these two observations, one immediately realizes that for a fixed value of ϵ_F and a given number of N_c, the warping effects become stronger with decreasing R, leading to stronger pseudo-TRS breaking.

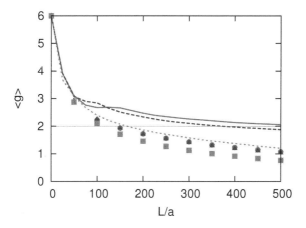

Fig. 3. Conductance of a disordered metallic carbon nanotube as a function of the length L of the disordered region measured in units of the lattice constant a. Solid; broken; dotted lines (filled circles; triangles; squares) correspond to the case without (with) dephasing, and of a different diameter of the nanotube: $M = 150$; $M = 100$; $M = 50$. The three cases are also represented by different colors: red; blue; gray. We set $\epsilon_F/t = 0.042$, 0.06309, and 0.12623, respectively to the above three cases so that the initial value of $\langle g \rangle$ always takes the same value: $\langle g \rangle_{L \to 0} = 2N_c = 6$. Other parameters are set as $W/t = 0.3$ and $p = 0.1$ and $L_\phi/a = 50$. The ensemble average is performed over 5000 samples. The magnitude of the error bar at $L/a = 500$ is of order 10^{-3}.

We assume that the potential profile of the scatterers is gaussian with its characteristic range d chosen to be $d/a = 1.5$, a value large enough for avoiding the inter-valley scattering. We then let the amplitude of this gaussian random potential w be uniformly (randomly) distributed within the range of $|w| \leq W/2$. As we mentioned earlier, the effects of dephasing has been taken account of by the approach employed also in Ref. 5, i.e., by separating the entire sample into several segments of equal length L_ϕ.[c]

Let us now look at Fig. 2. The main panel shows a linear plot of $\langle g \rangle$, indicating that $\langle g \rangle$ converges to unity irrespective of the presence or absence of dephasing; a clear signature of the appearance of a PCC. This partly confirms numerically our earlier prediction based on a Boltzmann equation approach, stating that "the PCC in a GNR is so robust that it may possibly survive even into the incoherent regime."[12] In our plots one can also observe that $\langle g \rangle$ in the presence of dephasing is slightly larger than the case of no dephasing. This feature is more clearly highlighted in the semilog plot of $\langle g \rangle - 1$. When $L/a \gtrsim 10^4$ (i.e., L is very large), the value of $\langle g \rangle - 1$ without dephasing scales away from a quasi-linear (stable) behavior in the presence of dephasing. This is probably due to residual inter-valley scattering.

---

[c]The rule of this game is the following: each time the incident electron leaves a segment and enters the next one, he loses his phase memory. As for concrete implementation of this to realistic carbon nanostructures, we refer interested readers to our forthcoming publication.

Notice that here dephasing plays indeed the role of *stabilizing the PCC* against weak inter-valley scattering.

Let us finally analyze our numerical data for CNTs (Fig. 3). We make a few remarks on our CNT data, which show a number of contrasting features to the case of GNR. First, the value of $\langle g \rangle$ is *smaller* in the presence of dephasing than in the absence of dephasing, which is consistent with the result of Ref. 5. This simply opposes the GNR case. In some cases ($M = 50$ and 100) $\langle g \rangle$ decreases even below the "protected" value of 2 as L/a increases. As mentioned earlier, trigonal warping of the Dirac cone is omnipresent whenever the Fermi level is away from the Dirac point, and this can possibly come into play in the transport characteristics of a CNT,[18] when its diameter or M is not large enough. This seemingly weak effect associated with the breaking of pseudo-TRS is shown to give a destructive influence on the scaling behavior of $\langle g \rangle$ in the large-L/a limit. Dephasing does not help. These observations lead us to our second conclusion that a stable existence of the two PCCs in a CNT is restricted to the case of a very *large diameter* and of a relatively *small doping*.

Acknowledgments

This work was supported in part by a Grant-in-Aid for Scientific Research (C) (No. 21540389) from the Japan Society for the Promotion of Science, and by the National Science Foundation under Grant No. NSF PHY05-51164.

References

1. See, for example, Y. Imry, *Introduction to Mesoscopic Physics*, 2nd edn. (Oxford University Press, Oxford, 2002).
2. T. Ando and T. Nakanishi, *J. Phys. Soc. Jpn.* **67**, 1704 (1998).
3. T. Ando, T. Nakanishi, and R. Saito, *J. Phys. Soc. Jpn.* **67**, 2857 (1998).
4. T. Nakanishi and T. Ando, *J. Phys. Soc. Jpn.* **68**, 561 (1999).
5. T. Ando and H. Suzuura, *J. Phys. Soc. Jpn.* **71**, 2753 (2002).
6. K. Wakabayashi, Y. Takane, and M. Sigrist, *Phys. Rev. Lett.* **99**, 036601 (2007).
7. K. Wakabayashi, Y. Takane, M. Yamamoto, and M. Sigrist, *Carbon* **47**, 124 (2009).
8. K. Wakabayashi, Y. Takane, M. Yamamoto, and M. Sigrist, *New J. Phys.* **11**, 095016 (2009).
9. M. Fujita, K. Wakabayashi, K. Nakada, and K. Kusakabe, *J. Phys. Soc. Jpn.* **65**, 1920 (1996).
10. C. Barnes, B. L. Johnson, and G. Kirczenow, *Phys. Rev. Lett.* **70**, 1159 (1993).
11. K. Hirose, T. Ohtsuki, and K. Slevin, *Physica E* **40**, 1677 (2008).
12. Y. Takane, *J. Phys. Soc. Jpn.* **79**, 024711 (2010).
13. Y. Takane, *J. Phys. Soc. Jpn.* **73**, 9 (2004).
14. Y. Takane, *J. Phys. Soc. Jpn.* **73**, 1430 (2004).
15. H. Sakai and Y. Takane: *J. Phys. Soc. Jpn.* **75**, 054711 (2006).
16. M. R. Zirnbauer, *Phys. Rev. Lett.* **69**, 1584 (1992).
17. A. D. Mirlin, A. Müller-Groeling, and M. R. Zirnbauer, *Ann. Phys.* (New York) **236**, 325 (1994).
18. K. Akimoto and T. Ando, *J. Phys. Soc. Jpn.* **73**, 2194 (2004).

DIRECTION DEPENDENCE OF SPIN RELAXATION IN CONFINED TWO-DIMENSIONAL SYSTEMS

PAUL WENK

School of Engineering and Science, Jacobs University Bremen,
Bremen, 28759, Germany
p.wenk@jacobs-university.de

STEFAN KETTEMANN

School of Engineering and Science, Jacobs University Bremen,
Bremen 28759, Germany, and
Asia Pacific Center for Theoretical Physics and Division of Advanced Materials Science
Pohang University of Science and Technology (POSTECH)
San31, Hyoja-dong, Nam-gu,
Pohang 790-784, South Korea
s.kettemann@jacobs-university.de

The dependence of spin relaxation on the direction of the quantum wire under Rashba and Dresselhaus (linear and cubic) spin-orbit coupling is studied. Comprising the dimensional reduction of the wire in the diffusive regime, the lowest spin relaxation and dephasing rates for (001) and (110) systems are found.

Keywords: spin orbit coupling; spin relaxation; quantum wire.

PACS numbers: 85.75.-d, 73.20.Fz, 75.76.+j, 75.40.Gb

1. Introduction

In a system with only Rashba spin-orbit coupling (RSOC) or linear Dresselhaus spin-orbit coupling (DSOC) in a (111) quantum well (QW) it is clear from the respective vector fields that a rotation in the plain should not effect the physics, i.e. the minimal spin relaxation rate, $1/\tau_s$, will not reduce. The linear RSOC does not depend on the growth direction at all. It is already known that on the other hand in a (001) 2D system with both bulk inversion asymmetry (BIA)[1] and structural inversion asymmetry (SIA)[2] we get an anisotropic spin relaxation. This has also been studied numerically in quasi-1D GaAs wires.[3] In this proceedings we present analytical results concerning this anisotropy for the 2D case as well as the case of QW with spin and charge conserving boundaries. Hereby we focus on materials where the dominant mechanism for spin relaxation is governed by the D'yakonov-Perel spin relaxation.

Furthermore, searching for long spin decoherence times at room temperature, the (110) QW attracted attention. Using the Cooperon Hamiltonian to find the spin relaxation times, the findings for systems with different growth directions can also be related to weak localization (WL) measurements.[4,5] Including the mentioned SOC effects, our Hamiltonian has the following form (we set $\hbar \equiv 1$)

$$H = \frac{1}{2m_e}\mathbf{k}^2 + V(\mathbf{x}) - \frac{1}{2}\gamma_g \boldsymbol{\sigma}\left(\mathbf{B}_{\text{SO}}(\mathbf{k})\right), \qquad (1)$$

where m_e is the effective electron mass, $\mathbf{B}_{\text{SO}}^T = (B_{\text{SO}x}, B_{\text{SO}y})$ the momentum dependent SO field and $\boldsymbol{\sigma}$ a vector, with components σ_i, $i = x, y, z$, the Pauli matrices. γ_g is the gyromagnetic ratio with $\gamma_g = g\mu_B$ with the effective g factor of the material, and $\mu_B = e/2m_e$ is the Bohr magneton constant. To analyze the spin relaxation for different wire directions we use for the SO interaction, which is caused by BIA, to lowest order in the wave vector $\mathbf{k}$ the general form,

$$-\frac{1}{2}\gamma_g \mathbf{B}_{\text{SO,D}}(\mathbf{k}) = \gamma_D \sum_i \hat{e}_i k_i (k_{i+1}^2 - k_{i+2}^2) \qquad (2)$$

where the principal crystal axes are given by $i \in \{x, y, z\}$, i cyclic, and the spin-orbit coefficient for the bulk semiconductor γ_D. Impurities are included through $V(\mathbf{x})$, which are uncorrelated and weak. To address both at the same time, the WL corrections as well as the spin relaxation rates in the system, our starting point is the Cooperon[6]. Maintaining time-reversal symmetry, we can relate the eigenvalues of the Cooperon to the spin relaxation rates. In the diffusive limit one finds the according Cooperon Hamiltonian H_c to be

$$H_c(\mathbf{Q}) := \frac{\hat{C}^{-1}(\mathbf{Q})}{D_e} = (\mathbf{Q} + 2e\mathbf{A_S})^2 + (m_e^2 E_F \gamma_D)^2 (S_x^2 + S_y^2), \qquad (3)$$

with the total momentum $\mathbf{Q} = \mathbf{p} + \mathbf{p}'$ which is the sum of the momenta of retarded and advanced propagating electron and accordingly $\mathbf{S} = (1/2)(\boldsymbol{\sigma} + \boldsymbol{\sigma}')$. The effective vector potential due to SO interaction is $\mathbf{A_S} = m_e \hat{\alpha} \mathbf{S}/e$, where $\hat{\alpha} = \langle \hat{a} \rangle$ is averaged over angle. The last term is due to appearance of cubic DSOC.

2. Spin Relaxation anisotropy in the (001) system

2.1. *2D system*

We rotate the system in-plane through the angle θ (the angle $\theta = \pi/4$ is equivalent to [110]). This does not affect the Rashba term but changes the Dresselhaus one to[7,8]

$$\frac{1}{\gamma_D}H_{D[001]}(\mathbf{k}) = \sigma_y k_y \cos(2\theta)(\langle k_z^2\rangle - k_x^2) - \sigma_x k_x \cos(2\theta)(\langle k_z^2\rangle - k_y^2)$$
$$- \sigma_y k_x \frac{1}{2}\sin(2\theta)(k_x^2 - k_y^2 - 2\langle k_z^2\rangle) + \sigma_x k_y \frac{1}{2}\sin(2\theta)(k_x^2 - k_y^2 + 2\langle k_z^2\rangle). \qquad (4)$$

The resulting Cooperon Hamiltonian, including both Rashba and DSOC, reads then

$$H_c(\mathbf{Q}) = (Q_x + \alpha_{x1}S_x + (\alpha_{x2} - q_2)S_y)^2 + (Q_y + (\alpha_{x2} + q_2)S_x - \alpha_{x1}S_y)^2$$
$$+ \frac{q_{s3}^2}{2}(S_x^2 + S_y^2), \qquad (5)$$

where we set $\frac{q_{s3}^2}{2} = (m_e^2 E_F \gamma_D)^2$, (6)

$$\alpha_{x1} = \frac{1}{2} m_e \gamma_D \cos(2\theta)((m_e v)^2 - 4\langle k_z^2 \rangle), \tag{7}$$

$$\alpha_{x2} = -\frac{1}{2} m_e \gamma_D \sin(2\theta)((m_e v)^2 - 4\langle k_z^2 \rangle) = \left(q_1 - \sqrt{\frac{q_{s3}^2}{2}}\right) \sin(2\theta)$$

$$= 2 m_e \tilde{\alpha}_1 \sin(2\theta), \tag{8}$$

with $q_1 = 2m_e\alpha_1$, $q_2 = 2m_e\alpha_2$. We see that the part of the Hamiltonian which cannot be written as a vector field and is due to cubic DSOC does not depend on the wire direction in the (001) plane.

2.2. *Quasi-1D wire*

In the following we consider spin and charge conserving boundaries. Due to the SOC we have to deal with modified Neumann boundary condition:

$$\left(-\frac{\tau}{D_e}\mathbf{n} \cdot \langle \mathbf{v}_F[\gamma_g \mathbf{B}_{SO}(\mathbf{k}) \cdot \mathbf{S}]\rangle - i\partial_\mathbf{n}\right) C|_{\partial S} = 0, \tag{9}$$

where $\langle ... \rangle$ denotes the average over the direction of Fermi velocity $\mathbf{v}_F$ and $\mathbf{k}$ and $\mathbf{n}$ is the unit vector normal to the boundary ∂S. For a system confined to the $x-y$ plane with the x-coordinate along the wire we have

$$(-i\partial_y + 2e(\mathbf{A_S})_y)C\left(x, y = \pm\frac{W}{2}\right) = 0, \quad \forall x, \tag{10}$$

with $A_y = (\alpha_{x2} + q_2)S_x - \alpha_{x1}S_y$ and $q_s = \sqrt{(\alpha_{x2} + q_2)^2 + \alpha_{x1}^2}$.

2.2.1. *Spin relaxation*

Searching for longest spin relaxations, we focus on systems with widths smaller than the spin precession length, which is motivated by the effect of motional narrowing.[9,10] We diagonalize the Hamiltonian, Eq. (5) with the boundary condition Eq. (10) using the scheme presented in Ref.9 wire width $Wq_s < 1$. Assuming the term proportional to W^2/k_x to be small, the absolute minimum can be found at

$$E_{min1} = \frac{3}{2}\frac{q_{s3}^2}{2} + \frac{\left(q_{sm}^2 - \frac{q_{s3}^2}{2}\right)\left(\alpha_{x1}^2 + \alpha_{x2}^2 - q_2^2\right)^2}{24 q_{sm}^2} W^2 \tag{11}$$

with $q_{sm} = \sqrt{(\alpha_{x2} - q_2)^2 + \alpha_{x1}^2}$. This minimum is independent of the width W if $\alpha_{x1}(\theta = 0) = -q_2$ and/or the direction of the wire is pointing in

$$\theta = \frac{1}{2} \arcsin\left(\frac{2\langle k_z^2 \rangle (m_e \gamma_D)^2 ((m_e v)^2 - 2\langle k_z^2 \rangle) - q_2^2}{(m_e^3 v^2 \gamma_D - 4\langle k_z^2 \rangle m_e \gamma_D) q_2}\right). \tag{12}$$

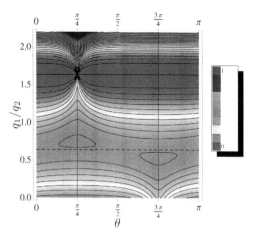

Fig. 1. Dependence of the W^2 coefficient in Eq. (13) on the lateral rotation (θ). The absolute minimum is found for $\alpha_{x1}(\theta = 0) = -q_2$ (here: $q_1/q_2 = 1.63$) and for different SO strength we find the minimum at $\theta = (1/4 + n)\pi$, $n \in \mathbb{Z}$ if $q_1 < (q_{s3}/\sqrt{2})$ (dashed line: $q_1 = (q_{s3}/\sqrt{2})$) and at $\theta = (3/4 + n)\pi$, $n \in \mathbb{Z}$ else. Here we set $q_{s3} = 0.9$. The scaling is arbitrary.

The second possible absolute minimum, which dominates for sufficient small width W and $q_{sm} \neq 0$, is found at

$$E_{min2} = \frac{q_{s3}^2}{2} + \frac{\left(\frac{q_{s3}^2}{2} + q_{sm}^2\right)\left(\alpha_{x1}^2 + \alpha_{x2}^2 - q_2^2\right)^2}{12 q_{sm}^2} W^2. \quad (13)$$

The minimal spin relaxation rate is found by analyzing the prefactor of W^2 in Eq. (13), as presented in Fig. (1). We see immediately that in the case of vanishing cubic DSOC or in the case where $\alpha_{x1}(\theta = 0) = -q_2$ we have no direction dependence of the minimal spin relaxation. Notice the shift of the absolute minimum away from $q_1 = q_2$ due to $q_{s3} \neq 0$. In the case of $q_1 < (q_{s3}/\sqrt{2})$ we find the minimum at $\theta = (1/4 + n)\pi$, $n \in \mathbb{Z}$, else at $\theta = (3/4 + n)\pi$, $n \in \mathbb{Z}$, which is indicated by the dashed line in Fig. (1).

2.2.2. *Spin dephasing*

Concerning spintronic devices it is interesting to know how an ensemble of spins initially oriented along the [001] direction dephases in the wire of different orientation θ. To do this analysis we only have to know that the eigenvector for the H_c eigenvalue

$$E(k_x = 0) = q_{s3}^2 + q_{sm}^2 - \frac{(\alpha_{x1}^2 + \alpha_{x2}^2 - q_2^2)^2 + \frac{q_{s3}^2}{2} q_s^2}{12} W^2, \quad (14)$$

corresponds to the to the z-component of the spin density whose evolution is described by the spin diffusion equation.[10] As an example we assume the case where cubic DSOC term can be neglected and where RSOC and lin. DSOC strength are

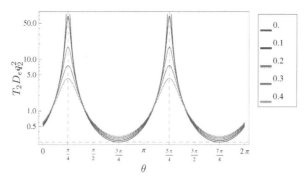

Fig. 2. The spin dephasing time T_2 of a spin initially oriented along the [001] direction in units of $(D_e q_2^2)$ for the special case of equal RSOC and lin. DSOC strength. The different curves show different strength of cubic DSOC in units of q_{s3}/q_2. In the case of finite cubic DSOC we set $W = 0.4/q_2$. If $q_{s3} = 0$, T_2 diverges at $\theta = (1/4 + n)\pi$, $n \in \mathbb{Z}$ (dashed vertical lines). The horizontal dashed line indicated the 2D spin dephasing time, $T_2 = 1/(4q_2^2 D_e)$.

equal. We notice that the dephasing is then width independent. At definite angles the dephasing time diverges,

$$\frac{1}{T_2(\theta)} = 2D_e q_2^2 (1 - \sin(2\theta)) \quad (15)$$

which is plotted in Fig. (2). We have longest spin dephasing time at $\theta = (1/4 + n)\pi$, $n \in \mathbb{Z}$. For $\theta = (3/4 + n)\pi$, $n \in \mathbb{Z}$ we get the 2D result $T_2 = 1/(4q_2^2 D_e)$. This gives an analytical description of numerical calculation done by J.Liu et al., Ref.3. Switching on cubic DSOC leads to finite spin dephasing time for all angles θ. In addition, T_2 is then width dependent. In the case of strong cubic DSOC where $q_{s3}^2/2 = q_1^2 = q_2^2$, the dephasing time T_2 is angle independent and for $q_{s3}^2/2 > q_1^2 = q_2^2$ the minima in $T_2(\theta)$ change to maxima and vice versa.

3. Spin relaxation in quasi-1D wire with [110] growth direction

To get the spin relaxation in a [110] QW with RSOC and DSOC we proceed as before and rotate the spacial coordinate system of the Dresselhaus Hamiltonian Eq.(2). Including RSOC (q_2), we end up with the following Cooperon Hamiltonian

$$\frac{C^{-1}}{D_e} = (Q_x - \tilde{q}_1 S_z - q_2 S_y)^2 + (Q_y + q_2 S_x)^2 + \frac{\tilde{q}_3^2}{2} S_z^2, \quad (16)$$

with $\tilde{q}_1 = 2m_e(\gamma_D/2)\langle k_z^2 \rangle - (\gamma_D/2)(m_e E_F/2)$, $q_2 = 2m_e \alpha_2$, and $\tilde{q}_3 = 3m_e E_F^2(\gamma_D/2)$. We see immediately that in the 2D case states polarized in the z-direction have vanishing spin relaxation as long as we have no RSOC. Compared with the (001) system the constant term due to cubic Dresselhaus does not mix spin directions. The appropriate Neumann boundary condition has now changed to $(-i\partial_y + 2m_e \alpha_2 S_x)C(x, y = \pm W/2) = 0$.

3.1. Lowest Spin Relaxation Rates

In the case without cubic DSOC the lowest spin relaxation rate is found at finite wave vectors $k_{x\ \text{min}} = \pm \Delta(1 - (q_2 W)^2/24)$,

$$\frac{1}{D_e \tau_s} = \frac{\Delta^2}{24}(q_2 W)^2, \text{ with } \Delta = \sqrt{\tilde{q}_1^2 + q_2^2}. \quad (17)$$

If cubic DSOC cannot be neglected, the absolute minimum of spin relaxation can also shift to $k_{x\ \text{min}} = 0$. This depends on the ratio of RSOC and lin. DSOC strength: If $q_2/q_1 \ll 1$, we find the absolute minimum at $k_{x\ \text{min}} = 0$,

$$E_{min1} = \frac{\tilde{q}_3 + \tilde{q}_1^2 + q_2^2}{2} - \Delta_c + \frac{1}{12}\Delta_c(q_2 W)^2, \quad (18)$$

$$\text{with } \Delta_c = \frac{1}{2}\sqrt{(\tilde{q}_3 + \tilde{q}_1^2)^2 + 2(\tilde{q}_1^2 - \tilde{q}_3)q_2^2 + q_2^4}. \quad (19)$$

If $q_2/q_1 \gg 1$, the absolute minimum shifts to $k_{x\ \text{min}} \approx \pm\Delta(1-(q_2W)^2/24)$. However, reducing wire width W will not cancel the contribution due to cubic DSOC to the spin relaxation rate.

4. Conclusions

Summarizing the results, we have characterized the anisotropy and width dependence of spin relaxation in a (001) QW. At [110] we find the longest spin dephasing time T_2. If the absolute minimum of spin relaxation is found at [110] or [$\bar{1}$10] direction depends on the strength of cubic Dresselhaus and wire width. The findings for the spin dephasing time are in agreement with numerical results. The analytical expression for T_2 allows to see directly the interplay between the cubic DSOC and the dimensional reduction, having effect on T_2. In addition, we analyzed the special case of a (110) system and found the minimal spin relaxation rates depending on Rashba and lin. and cubic DSOC in the presence of boundaries. These results can be used to understand width and direction dependent WL measurements in QWs.

Acknowledgments

This research was supported by DFG-Project KE 807/6-1 and by the National Research Foundation of Korea through R31-2008-000-10059-0, Division of Advanced Materials Science.

References

1. G. Dresselhaus, *Phys. Rev.* **100**, 580 (1955).
2. Y. Bychkov and E. Rashba, *JETP Lett.* **39**, 78 (1984).
3. J. Liu, T. Last, E. Koop, S. Denega, B. van Wees, and C. van der Wal, *J. Supercond. Nov. Magn.* **23**, 11 (2010).
4. P. Wenk and S. Kettemann, *Phys. Rev. B* **83**, 115301 (2011).

5. M. Scheid, M. Kohda, Y. Kunihashi, K. Richter, and J. Nitta, *Phys. Rev. Lett.* **101**, 266401 (2008).
6. S. Hikami, A. I. Larkin, and Y. Nagaoka, *Prog. Theor. Phys.* **63**, 707 (1980).
7. J. Cheng, M. Wu, and I. da Cunha Lima, *Phys. Rev. B* **75**, 205328 (2007).
8. M. W. Wu, J. H. Jiang, and M. Q. Weng, *Physics Reports* **493**, 61 (2010).
9. S. Kettemann, *Phys. Rev. Lett.* **98**, 176808 (2007).
10. P. Wenk and S. Kettemann, *Phys. Rev. B* **81**, 125309 (2010).

ANALYSIS OF QUANTUM CORRECTIONS TO CONDUCTIVITY AND THERMOPOWER IN GRAPHENE — NUMERICAL AND ANALYTICAL APPROACHES

ALEKSANDER P. HINZ and STEFAN KETTEMANN

School of Engineering and Science, Jacobs University Bremen, 28759 Bremen, Germany, and Division of Advanced Materials Science, Pohang University of Science and Technology (POSTECH), San31, Hyoja-dong, Nam-gu, Pohang 790-784, South Korea
ahinz@postech.ac.kr & s.kettemann@jacobs-university.de

EDUARDO R. MUCCIOLO

Department of Physics, University of Central Florida, Orlando, Florida 32816, USA
mucciolo@physics.ucf.edu

We present numerical and analytical studies of the crossover between weak antilocalization and weak localization in monolayer graphene and their influence on thermopower. By the use of the recursive Green's function method, we find that these quantum corrections result in an enhancement of thermopower, which can be observed in the resulting magnetic field dependence. This magneto thermopower strongly depends on the size and strength of the impurities as well as on the back gate voltage of the system and the impurity concentration. We show in detail the crossover of these localization effects with these parameters. Using the disorder parameters of the numerical calculation, we find quantitative agreement with the analytical calculations.

Keywords: Weak localization; graphene; thermopower.

PACS numbers: 65.80.Ck, 72.20.Pa, 73.22.Pr

1. Introduction

We present numerical and analytical investigations of the influence of weak localization (WL) and weak antilocalization (WAL) on electrical conductivity and thermopower in graphene with varying disorder and system parameters.

2. Numerical method

The numerical calculations employ the recursive Green's function method[1]. A tight-binding model of electrons hopping between nearest neighbors in a honeycomb lattice is used, with hopping amplitude $t = 2.7$ eV. Contacts to leads are placed on opposite sides along zigzag edges. The leads are modeled by semi-infinite square lattices with propagating modes matching the honeycomb structure. Impurities are randomly distributed over the honeycomb lattice. Different realizations of disorder

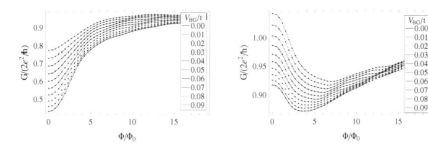

(a) WL for several V_{BG}, $K_0 = 0.5$.

(b) WAL for several V_{BG}, $K_0 = 4$.

Fig. 1. Magnetoconductance. Φ is the total magnetic flux through the sample in units of the magnetic flux quantum Φ_0, $W/L = 1$, $\xi/a_0 = 4$, $n_{imp} = 0.03$. System size is $L = 20a_0$.

correspond to different impurity distributions. A magnetic field is introduced via Peierls substitution in the hopping amplitudes.

3. Numerical results I — Specific systems

Assuming that disorder is due to impurities with a Gaussian profile, we can tune WL and WAL effects in a number of ways. The impurity potential range ξ, its strength V or its density n_{imp} can all be varied individually or together, as in the dimensionless number K_0 defined from the impurity potential correlation function

$$\langle V(r_i) V(r_j) \rangle = \frac{K_0(\hbar v_0)^2}{2\pi \xi^2} e^{\frac{-|r_i-r_j|^2}{2\xi^2}}, \text{ with } V(r_j) = \sum_{n=1}^{N_{imp}} V_n e^{-\frac{|r_j-R_n|^2}{2\xi^2}} \quad (1)$$

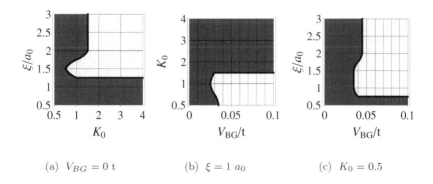

(a) $V_{BG} = 0$ t.

(b) $\xi = 1\, a_0$.

(c) $K_0 = 0.5$.

Fig. 2. Two-dimensional localization phase diagrams. Positive $d\Delta\sigma(B)/dB$ for small magnetic field is shown in blue (WL); negative $d\Delta\sigma(B)/dB$ is shown in white (WAL), $d\Delta\sigma(B)/dB$ is set by (5). Calculated points are indicated by grid points. The black line indicates the WL/WAL crossing.

where v_0 is the Fermi velocity, N_{imp} is the total number of impurities, R_n are the impurity positions and $V_n \in [-\delta, \delta]$ is the strength of the n-th impurity. At high doping and in the regime of dilute impurities, K_0 is proportional to $1/(k_F l)$, where k_F is the Fermi wavenumber and l is the elastic mean free path. Therefore, K_0 is a useful dimensionless measure of the disorder strength. The system's linear size for all results displayed is $L = 30a_0$, if not written differently, where a_0 is the lattice constant, W is the width and L is the length of the sample, $W/L = 1$ and $n_{imp} = 0.03$. The transition is also observed when changing the back gate voltage V_{BG}. An example of the WL and WAL dependence of the conductance on the gate voltage is shown in Fig.(1), which is consistent with experimental observations[2,3,4].

4. Theory — Localization effects in graphene

Electronic states in the disordered honeycomb lattice can be described conveniently by introducing an isospin $\vec{\Xi}$ (to account for the sublattice structure) and a pseudospin $\vec{\Lambda}$ (to account for the two subvalleys) into the Hamiltonian (see Ref. [5]),

$$H = \underbrace{v_F \vec{\Xi} \vec{k}}_{Dirac\ cone} - \underbrace{\mu \Xi_1 (\vec{\Xi} \vec{k}) \Lambda_3 \Xi_1 (\vec{\Xi} \vec{k}) \Xi_1}_{First\ correction\ to\ the\ Dirac\ cone} + \underbrace{I * V_{0,0}(\vec{r}) + \sum_{i,j=1}^{3} \Xi_i \Lambda_j V_{i,j}(\vec{r})}_{Impurity} \quad (2)$$

The total weak localization amplitude emerges from the interplay of the following four effects:

Contrapropagating electrons that are scattered on closed travel paths interfere at the loop starting points constructively due to the equal phase change on each path, leading to an enhancement of backscattering and thereby a tendency towards WL. The isospin is strongly coupled to the momentum of the electron, so that the isospin relaxation rate equals the momentum scattering rate, resulting in the disappearance of the isospin triplet terms and thereby a preference for WAL.

On the other hand, the intervalley scattering mixes valleys, which is accompanied by a change of chirality. This leads to a pseudospin relaxation and therefore tends to alter the conductance to WL.

Finally, intravalley scattering (long distance scatterers) leads to a suppression of $m = \pm 1$ pseudospin triplet terms which prefers WAL.

All these effects add up to yield the total conductivity correction $\Delta\sigma(B) = \sigma(B) - \sigma(B = 0)$, where[5]

$$\frac{\Delta\sigma(B)}{e^2/(\pi h)} = \left(\overbrace{F\left(\frac{\tau_B^{-1}}{\tau_\phi^{-1}}\right)}^{PS\ singlet} - \underbrace{F\left(\frac{\tau_B^{-1}}{\tau_\phi^{-1} + 2\tau_i^{-1}}\right)}_{Isospin\ singlet} - \overbrace{2 * F\left(\frac{\tau_B^{-1}}{\tau_\phi^{-1} + \tau_i^{-1} + \tau_*^{-1}}\right)}^{PS\ triplet} \right) \quad (3)$$

$$\propto \left(1 - (1 + 2(\tau_\phi/\tau_i))^{-2} - 2(1 + \tau_\phi/\tau_*)^{-2}\right) B \quad \text{for small B.} \quad (4)$$

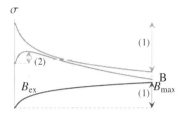

Fig. 3. Sketch of the amplitude $\Delta\sigma_{amp}$ defined in Eq. (6). For pure WAL and WL, the amplitude is set relative to σ at a magnetic field where the localization correction vanishes ($B = B_{max}$), cases (1), or else where the localization correction maximum shows a maximum $B = B_{ex}$, case (2).

Here, $\tau_B^{-1} = 4eDB/\hbar$, $\tau_*^{-1} = \tau_w^{-1} + \tau_z^{-1}$ (w denotes sheet warping), $F(z) = \ln(z) + \psi\left(0.5 + z^{-1}\right)$, ψ is the digamma function, τ_ϕ^{-1} is the dephasing rate, $1/\tau_z$ the intravalley scattering rate, $1/\tau_i$ the intervalley scattering rate and D is the diffusion constant.

5. Numerical results II — Conductivity phase diagrams

In order to characterize the nature of the localization regime, the sign of the slope for $B \to 0$ of the magnetoconductivity is computed from

$$d\Delta\sigma(B)/dB \sim \sigma(B \to 0) - \sigma(B = 0). \tag{5}$$

By varying two impurity parameters and keeping a third fixed, we generate two-dimensional phase diagrams as in Fig.(2). When we vary three parameters at the same time, we obtain the three-dimensional plots shown in Fig.(4(a)). In addition, we calculate the amplitude of magnetoconductivity by evaluating its quantity via:

- For weak localization

$$\Delta\sigma_{amp}(B) = \begin{cases} \sigma(B=0) - \sigma(B = B_{ex}) & \text{if localization correction} \\ & \text{shows an extremum} \\ \sigma(B=0) - \sigma(B = B_{max}) & \text{otherwise} \end{cases}$$

- For weak antilocalization

$$\Delta\sigma_{amp}(B) = \sigma(B=0) - \sigma(B = B_{max}), \tag{6}$$

where B_{max} is the magnetic field at which the localization correction disappears. With this quantity, displayed in Fig.(3) we can introduce additional contour lines into the three-dimensional phase diagrams ($\Delta\sigma(B) = 0$ line from Fig.(2) is consistent with the $\Delta\sigma_{amp}(B) = 0$ case), which we then name continuous diagrams, see Fig.(4(b)).

Crossovers between distinct localization regimes are observed - for instance when the impurity range shifts from short ($\xi < a_0$, WL) to long ($\xi > a_0$, WAL) or when $K_0 \approx 1$ (visible in Fig.(4(a))) is close to the Dirac point.

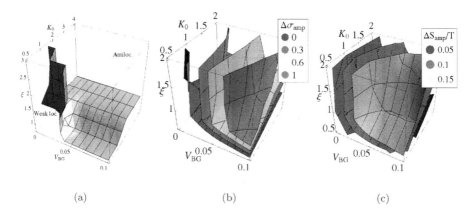

Fig. 4. (a) 3d conductivity phase diagram obtained with Eq. (5) merging the three diagrams from Fig.(2). (b) Conductivity amplitude $\Delta\sigma_{amp}$ phase diagram obtained with Eq. (6). (c) Thermopower amplitude phase diagram with Eq. (8). $\Delta\sigma_{amp}$ is in units of $2e^2/h$ while $\Delta S_{amp}/T$ is in units of $\mu V/K^2$. Each grid line crossing point corresponds to a numerically calculated value.

6. Numerical results III — Thermopower

We calculate thermopower via the Mott formula, which is valid for small temperature and large chemical potential or gate voltage. It is given by

$$S = -\frac{\pi^2 k_B^2 T}{3e} \frac{1}{\sigma} \frac{d\sigma}{dV_{\text{BG}}} \frac{dV_{\text{BG}}}{dE}\bigg|_{E=E_F}. \qquad (7)$$

For some impurity parameters we find an enhancement of thermopower due to localization [see Fig.(5)]. It is manifest in three-dimensional diagrams analogously to those obtained for $\Delta\sigma_{amp}$, as can be seen in Fig.(4(c)). There, the quantity presented is $\Delta S_{amp}/T$, in which ΔS_{amp} is defined by

$$\Delta S_{amp} = S(B=0) - S_{mean}, \qquad (8)$$

where S_{mean} is the average value of thermopower at high magnetic field. By comparing Fig.(4(b)) to Fig.(4(c)), we find that the enhancement of thermopower is directly related to the strength of WAL, since the derivative of the electrical conductivity with respect to back gate voltage depends solely on the isospin singlet pseudospin triplet terms, which are also responsible for WAL.

7. Analytical results

We use the connection between the impurity potential strength and the scattering rate, $\tau^{-1} = \left(\pi\gamma\langle V^2\rangle\right)/\hbar$, where γ is the density of state, together with Eq. (3) to obtain phase diagrams for both conductivity and thermopower analytically. As a

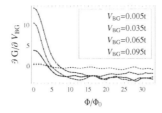

Fig. 5. The numerically calculated factor $\frac{dG}{dV_{BG}}$ responsible for the enhancement in thermopower in the Mott formula (7). $t = 2.7 eV$, G is the conductance.

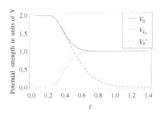

Fig. 6. Dependence of V_0, V_0' and V_{k_0} on ξ. ξ in units of a_0.

first result, we find

$$\frac{\tau_\phi}{\tau_i} = C_1 V_{k_0}^2 (V_0 + V_0')^2 n_{imp}^2 V_{BG}^2 \qquad (9)$$

$$\frac{\tau_\phi}{\tau_z} = C_2 (V_0 - V_0')^2 (V_0 + V_0')^2 n_{imp}^2 V_{BG}^2, \qquad (10)$$

where V_0 and V_0' are potential amplitudes related to sublattices A and B for intravalley scattering and V_{k_0} and V_{k_0}' are the related intervalley scattering amplitudes of the Fourier transformed impurity potential. While V_{k_0}' is always zero, the dependence of the other potential strengths on ξ is displayed in Fig.(6). C_1 and C_2 are numerical prefactors. For all calculations the sheet warping term is set to zero, so $\tau_* = \tau_z$.

In the case of conductivity, the WL-WAL crossover given by Eq. (4) can be displayed in a τ_ϕ/τ_z - τ_ϕ/τ_i - phase diagram, see Fig.(7). For instance, Eq. (10) allows us to identify the localization crossing line by tuning impurity parameters. Following this approach for thermopower, we find a positive correction due to localization only in the WAL regime, Fig.(7(b)), which is consistent with the numerical calculations.

8. Conclusion and outlook

By varying the impurity range and other system parameters, we observe numerically a weak localization/antilocalization crossing for the conductivity, which is consistent with the analysis of Ref. [5]. In addition, a thermopower enhancement in the region corresponding to WAL is discernible both in the numerical as well as in the

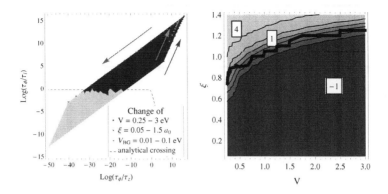

Fig. 7. *Left*: $d\Delta\sigma(B)/dB$ phase diagrams displayed in the $\tau_\phi/\tau_z - \tau_\phi/\tau_i$ parameter space. The WAL area is shown in yellow and the WL area is red. *Right*: $\frac{d\Delta\sigma(B=0)}{dV_{BG}}$ positive correction due to localization in units of $(2e^2/h)/V_{BG}$ indicated in yellow in the WAL area and suppression in red in the WL area. *Both*: Black line is the WL-WAL crossing line calculated via Eq. (4); $V_{BG} = 0.01 eV$.

analytical findings, resulting from the back gate voltage dependence of the pseudospin triplet term. In the future, we plan to reinvestigate this matter, replacing the Mott formula with the Kubo formula.

Acknowledgments

This research was supported by WCU (World Class University) program through the National Research Foundation of Korea funded by the Ministry of Education, Science and Technology(R31-2008-000-10059-0).

References

1. D. K. Ferry and S. M. Goodnick, *Transport in Nanostructures* (Cambridge University Press, 1977)
2. Tikhonenko, F. V. and Kozikov, A. A. and Savchenko, A. K. and Gorbachev, R. V., Phys. Rev. Lett. **103**, 226801 (2009)
3. Zuev, Y. M. and Chang, W. and Kim, P., Phys. Rev. Lett. **102**, 096807 (2009)
4. Ki, D. K. and Jeong, D. and Choi, J. H. and Lee, H. J., Phys. Rev. B **78**, 125409 (2008)
5. McCann, E. and Kechedzhi, K. and Fal'ko, Vladimir I. and Suzuura, H. and Ando, T. and Altshuler, B. L., Phys. Rev. Lett. **97**, 146805 (2006)

Localisation 2011
International Journal of Modern Physics: Conference Series
Vol. 11 (2012) 177–182
© World Scientific Publishing Company
DOI: 10.1142/S2010194512006095

INDIRECT EXCHANGE INTERACTIONS IN GRAPHENE

HYUNYONG LEE,[1,*] EDUARDO R. MUCCIOLO,[2,†] GEORGES BOUZERAR,[3,4,‡] and STEFAN KETTEMANN[1,4,§]

[1]*Division of Advanced Materials Science, Pohang University of Science and Technology (POSTECH), Pohang 790-784, South Korea*
[2]*Department of Physics, University of Central Florida, Orlando, Florida 32816, USA*
[3]*Institut Néel, CNRS, département MCBT, 25 avenue des Martyrs, BP 166, 38042 Grenoble Cedex 09, France*
[4]*School of Engineering and Science, Jacobs University Bremen, Bremen 28759, Germany*
*hyunyongrhee@postechedu
†mucciolo@physics.ucf.edu
‡georges.bouzerar@grenoble.cnrs.fr
§s.kettemann@jacobs-university.de

We study the Ruderman-Kittel-Katsuya-Yoshida (RKKY) interactions in graphene controlling the gate voltage and applying nonmagnetic disorder. It is found that oscillations of the RKKY interactions in undoped graphene are characterized by the interference of two neighbor Dirac nodes $\boldsymbol{K}$ and $\boldsymbol{K'}$ in the first Brillouin zone and decays with R^{-3} distance dependence. In the slightly doped graphene, a beating pattern, which consists of two characteristic wavevectors ($\boldsymbol{K} - \boldsymbol{K}$ and $\boldsymbol{k_F}$), starts to appear. The distance dependence in this regime shows a crossover from the R^{-3} to R^{-2}. We present the effect of weak disorder on the RKKY interactions in diffusive regime. The arithmetic averaged interaction over disorder configurations decreases exponentially at distances exceeding the elastic mean free path, while the geometrical average(typical) value has the same power-law as the clean limit.

Keywords: RKKY; graphene.

PACS number: 71.70.Gm

1. Introduction

Many analytical and numerical studies on the RKKY interactions in clean graphene at Dirac (neutrality) point have been reported recently.[1–4] Due to the particle-hole symmetry, in bipartite lattices including graphene, the RKKY interactions have always ferromagnetic correlation between the magnetic impurities on the same sublattice, but antiferromagnetic correlation between the ones on the different sublattice.[1] Because of the pseudo gap at Dirac point, the distance (R) dependence of the RKKY interactions in graphene is $1/R^3$, which is unexpected behavior in 2 dimensional metal systems ($1/R^2$). Due to the presence of the two Dirac points $\boldsymbol{K}^{(')} = (\pm 2\pi/3\sqrt{3}, 2\pi/3)$ the RKKY interaction is governed by two different length scales, $|\boldsymbol{K} - \boldsymbol{K'}|^{-1}$ and the Fermi wavelength k_F^{-1}, in doped (but near the Dirac point)

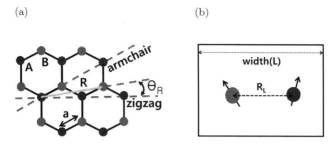

Fig. 1. (a) Schematic diagrams of graphene lattice. The two sublattices are denoted as A and B and the two representative directions (zigzag and armchair) are indicated as dashed gray lines. θ_R is the angle between the displacement vector of the magnetic impurity $\boldsymbol{R}$ and the zigzag direction. (b) A lattice setup with the system size L and the furthermost distance R_L between impurity spins.

graphene.[4] The on-site disorder scatters the phase of the electron's wave function and changes its amplitude randomly so that the period of the oscillation and its amplitude are altered comparing to the one in the clean system. In diffusive regime, consequently, the averaged RKKY interaction over disorder configurations decays exponentially on distances exceeding the mean free path scale l_e. However its geometrical average value has the same power law decaying as the clean sample[8] so that weak disorder is not likely to cause any critical change in physical properties which derive from the RKKY interaction. We summarize the results obtained recently.[8,9]

2. Model and method

We start with a general expression for the RKKY exchange coupling constant in terms of the unperturbed (disorder-free) electronic Green's function $G^{(0)}(\boldsymbol{r}_i, \boldsymbol{r}_j, \omega)$,[5]

$$J_{\text{RKKY}} = J^2 \frac{S(S+1)}{4\pi S^2} \int d\omega\, f(\omega)\, \text{Im}\left[G^{(0)}(\boldsymbol{r}_j, \boldsymbol{r}_i, \omega) \times G^{(0)}(\boldsymbol{r}_i, \boldsymbol{r}_j, \omega) \right] \quad (1)$$

$$= J^2 \frac{S(S+1)}{4\pi S^2} \text{Im} \int d\omega\, f(\omega) \times \sum_{n,m} \frac{F_{nm}^{ij}}{(E_n - \omega + i\delta)(E_m - \omega + i\delta)}. \quad (2)$$

where J is the local coupling constant between the localized magnetic impurities and the itinerant electrons, S is the magnitude of the impurity spin, i (j) is the site index of a magnetic impurity located at position $\boldsymbol{r}_i$ ($\boldsymbol{r}_j$), $f(\omega) = [e^{(\omega-\mu)/T} + 1]^{-1}$ is the Fermi-Dirac distribution function, and $F_{nm}^{ij} = \psi_n^*(\boldsymbol{r}_i)\psi_n(\boldsymbol{r}_j)\psi_m^*(\boldsymbol{r}_j)\psi_m(\boldsymbol{r}_i)$, with $\psi_n(\boldsymbol{r}_i)$ denoting the eigenfunction corresponding to the eigenenergy E_n of the electronic disorder-free Hamiltonian. The lattice constant a and the Planck's constant $\hbar$ are set to unity in all numerical calculations. Using a zero-temperature approximation ($T = 0$) and changing to an integral form, Eq. (2) can be recast as

$$J_{\text{RKKY}} = -J^2 \frac{S(S+1)}{2S^2} \int_{E<\mu} dE \int_{E'>\mu} dE' \frac{F(E, E')}{E - E'}, \quad (3)$$

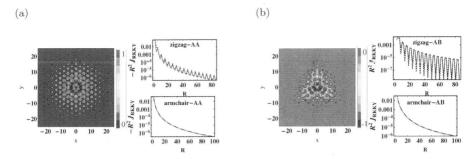

Fig. 2. Plots of the RKKY interaction strengths between a magnetic impurity at the origin and another at: (a) a site from the same sublattice (AA) and (b) a site from a different sublattice (AB). The amplitudes are multiplied by the square of the distance to facilitate visualization. The lattice constant is set to unity. The numerical data is for clean graphene ($W = 0$). Calculations using the kernel polynomial method and lattice Green's function method are represented as solid black and dashed red lines, respectively.

where $F(E, E') = \text{Re}[\rho_{ji}(E)\rho_{ij}(E')]$, μ is the Fermi energy, and $\rho_{ij}(E) = \langle i|\delta(E - H)|j\rangle$, which can be calculated numerically using the kernel polynomial method (KPM).[6,7] In the KPM, the matrix elements $\rho_{ij}(E)$ are expressed as sums over order-M Chebyshev polynomials on the energy E with coefficients obtained through an efficient recursion relation involving matrix elements of the system Hamiltonian. As our unperturbed electronic Hamiltonian with on-site disorder, we employ the single-band Anderson tight-binding model on a honeycomb lattice,

$$H = -t \sum_{\langle i,j \rangle} c_i^+ c_j + \sum_i w_i\, c_i^+ c_i, \qquad (4)$$

where t (≈ 2.67 eV for graphene) is the hopping energy, c_i (c_i^+) annihilates (creates) an electron at site i, w_i is the on-site random disorder energy distributed uniformly between $[-W/2, W/2]$, and $\langle i, j \rangle$ denote nearest-neighbor sites. Periodic boundary condition is used for all calculations.

3. Numerical results

3.1. *Clean system*

For pure systems ($W = 0$) at the neutrality point ($\mu = 0$), the Chebyshev polynomials are calculated up to $M = 3 \times 10^3$ on a lattice with 5×10^5 sites and the numerical results are shown in Fig. 2, where the R^2 are multiplied to emphasis the oscillatory behavior of J_{RKKY} resulting in a smoother ($1/R$) decay. The interactions along the zigzag and armchair directions are shown separately by line plots in Fig. 2. These results are in excellent agreement with previous studies.[1–3] The authors of Ref. 3

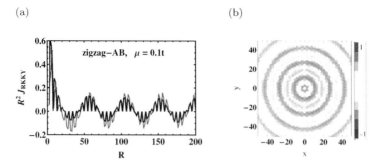

Fig. 3. The RKKY interaction multiplied by R^2 in doped graphene (a) $\mu = 0.1t$ along zigzag-AB, density plots for (b) $\mu = 0.3t$, where t is the nearest-neighbor hopping energy. The lattice constant is set to unity. Calculations using the kernel polynomial method and lattice Green's function method are represented as solid blue and dashed red line respectively in (a). A lattice with 7.2×10^5 and a polynomial degree cut off of $M = 5 \times 10^3$ are used in these numerical calculations. The lattice constant a is set to unity.

used a lattice Green's function method to obtain an RKKY interaction of the form

$$J_{AA}^0 = -J^2 \frac{1 + \cos[(\boldsymbol{K} - \boldsymbol{K}') \cdot \boldsymbol{R}]}{R^3}, \quad (5)$$

$$J_{AB}^0 = J^2 \frac{3 + 3\cos[(\boldsymbol{K} - \boldsymbol{K}') \cdot \boldsymbol{R} + \pi - 2\theta_R]}{R^3}, \quad (6)$$

where all the coefficients are set to unity, $\boldsymbol{R} = \mathbf{r}_i - \mathbf{r}_j$, and θ_R is defined in Fig. 1a. For a direct comparison, plots of Eqs. (5) and (6) are shown in Fig. 2 along with the results calculated from Eq. (3). As expected from the particle-hole symmetry of the spectrum, the magnetic impurity on the origin has ferromagnetic correlations with the impurities on the same sublattice (Fig. 2a), while antiferromagnetic correlations develop for impurities on different sublattices (Fig. 2b).

By controlling μ in Eq. (3), we investigate how the RKKY interaction evolves with the Fermi level and the results are shown in Fig. 3, where the R^2 are multiplied to emphasize the oscillatory behavior of J_{RKKY}. We have used a lattice with 7.2×10^5 and a polynomial degree cut off of $M = 3 \times 10^3$ in these KPM calculations. Near the Dirac point (Fig. 3a, b), beating pattern appears and it consists of the waves which are characterized by two wave vectors, $\boldsymbol{K} - \boldsymbol{K}'$ and $\boldsymbol{q}_F$, where $\boldsymbol{q}_F$ is the Fermi wave vector. With electron or hole doping the particle-hole symmetry is broken, so that it causes the sign alteration with the distance between the moments. Recently, for this beating behavior, analytical expression using the lattice Green's function has been reported,[4]

$$J_{AA} = J_{AA}^0 \left[1 + \frac{8 q_F R}{\sqrt{\pi}} G_{1,3}^{2,0} \left(\begin{matrix} \frac{1}{2}, \frac{3}{2} \\ 1, 1, 1 \end{matrix} ; q_F^2 R^2 \right) \right], \quad (7)$$

$$J_{AB} = J_{AB}^0 \left[1 - \frac{8 q_F R}{\sqrt{3\pi}} G_{2,4}^{2,1} \left(\begin{matrix} \frac{1}{2}, \frac{3}{2} \\ 1, 2, 0, -\frac{1}{2} \end{matrix} ; q_F^2 R^2 \right) \right], \quad (8)$$

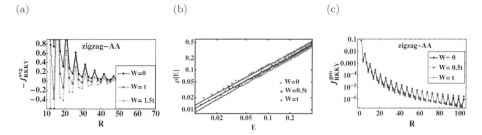

Fig. 4. Plots of the RKKY interaction strength along the zigzag in the diffusive regime, as (a) averaged and (b) geometrical averaged over 1600 different disorder configurations. A lattice with 1.8×10^5 sites and a polynomial degree cutoff of $M = 5 \times 10^3$ are used in these numerical calculations. (c) The averaged density of states in log scales, where a lattice with 3×10^7 sites and a polynomial degree cutoff of $M = 7 \times 10^3$ are used. The black lines in (c) represent fittings to the relation $\rho(E) = \gamma |E|$ with $\gamma = 0.95, 1.07, 1.3$ for $W = 0, 0.5t, t$, respectively

where $J^0_{AA(B)}$ is the coupling function at Dirac point [Eq. (5),(6)], G is the Meijer-G function. Note that the expression describing the oscillatory behavior with respect to the Fermi momentum q_F, depends only on its magnitude q_F not its direction, contrary to the $J^0_{AA(B)}$ which depends on both the direction and magnitude of the interference of the two neighbor Dirac point $\bm{K} - \bm{K'}$. To compare with our calculation, the result of Eq. (8) is also presented by black solid line in Fig. 3a and these are in excellent agreement each other.

3.2. Disordered system

In order to investigate the effect of on-site nonmagnetic disorder in graphene, we consider 1600 different disorder configurations for each value of W and then evaluate the matrix elements ρ_{ij} through the KPM with $M = 5 \times 10^3$ on a lattice with 1.8×10^5 sites. For weak disorder strength, the system is in the diffusive regime; the actual value of W where this crossover occurs depends on the lattice size and it has been determined by evaluating the localization length. The average amplitude of the RKKY interaction in the diffusive regime is shown in Fig. 4a. Similarly to conventional metals, the interaction decays exponentially with increasing disorder strength as $J^{\text{avg}}_{\text{RKKY}} \sim J^{\text{clean}}_{\text{RKKY}} e^{-R/l_e}$, where l_e is the mean free path and $J^{\text{clean}}_{\text{RKKY}}$ is the interaction amplitude in the clean limit. It is worth noticing that the sign of the interaction oscillates when the impurities are located along the zigzag AA direction.

To better characterize the amplitude of the interaction, we have also calculated the geometrical average ($J^{\text{geo}}_{\text{RKKY}}$) for diffusive regimes (Fig. 4b). We observed that the geometrical average for weakly disordered system has a decaying behavior similar to the clean system, as previous works expected.[8] We have calculated the density of state for two disordered system ($W = 0.5t, t$) to observe how the weak disorder affects the pseudo gap at the Dirac point ($E = 0$) which leads to the unconventional decaying behavior of the RKKY interaction ($1/R^3$). For the density of state calculation, the KPM has been also used with a 3×10^7 sites and a polynomial degree cutoff of $M = 7 \times 10^3$. As one may see from the density of state calculation (Fig. 4c), the pseudo gap, $\rho(E) = \gamma |E|$, is still not filled with weak disorder and only the slope (γ) around the Dirac point is changed, which means that the geometrical average of the RKKY interaction may not change drastically from the clean limit.

4. Conclusion

In conclusion, we have confirmed that the RKKY interactions in clean graphene has a strong anisotropy of its sign and oscillation amplitude, and it decays as $1/R^3$ for all directions. We also have studied the effect of the gate voltage and non magnetic disorder on the RKKY interaction with the KPM. The finite gate voltage breaks the particle-hole symmetry and gives the finite Fermi surface which yields the Friedel oscillation behavior in long rage limit, so that the sign of the RKKY interaction between the impurities localized on the same sublattice oscillates with distance. Increasing the strength of nonmagnetic, on-site disorder causes the averaged amplitude of the RKKY interaction to decrease exponentially at distances exceeding the elastic mean free path. With the accurate evaluation of the density of states around the neutrality point in weakly disordered regime ($W \leq t$), we have confirmed that the linear relation is still valid and the pseudo gap is not filled. Therefore, we may conclude that the geometrical average of the RKKY interaction in diffusive regime decays with $1/R^3$ as in the clean system, not $1/R^2$ as in the usual 2 dimensional metal.

Acknowledgments

This research was supported by WCU(World Class University) program through the National Research Foundation of Korea funded by the Ministry of Education, Science and Technology(R31-2008-000-10059-0)

References

1. S. Saremi, Phys. Rev. B **76**, 184430 (2007).
2. A. M. Black-Schaffer, Phys. Rev. B **81**, 205416 (2010).
3. M. Sherafati and S. Satpathy, Phys. Rev. B **83**, 165425 (2011).
4. M. Sherafati and S. Satpathy, Phys. Rev. B **84**, 125416 (2011).
5. R. M. White, *Quantum Theory of Magnetism* (McGraw-Hill, New York, 1970).
6. Roche, Stephan and Mayou, Didier, Phys. Rev. B. **60**, 322 (1999).
7. A. Weiße, G. Wellein, A. Alvermann, and G. Fehske, Rev. Mod. Phys. **78**, 275 (2006).
8. P. F. de Chatel, J. Magn. Magn. Mater. **23**, 28 (1981).
9. H. Lee, E. R. Mucciolo, G. Bouzerar and S. Kettemann, Phys. Rev. B 85, 075420 (2012).
10. H. Lee, E. R. Mucciolo, G. Bouzerar and S. Kettemann, unpublished.

CRITICAL EXPONENTS FOR ANTIFERROMAGNETIC SPIN CHAINS OBTAINED FROM BOSONISATION

MARCEL KOSSOW,[1,*] PETER SCHUPP[1,†] and STEFAN KETTEMANN[1,2,‡]

[1] *School of Engineering and Science, Jacobs University Bremen, Bremen 28759, Germany*
[2] *Division of Advanced Materials Science, Pohang University of Science and Technology (POSTECH), Pohang 790-784, South Korea*
m.kossow@jacobs-university.de
†*p.schupp@jacobs-university.de*
‡*s.kettemann@jacobs-university.de*

The Heisenberg spin 1/2 chain is revisited in the perturbative RG approach with special focus on the transition of the critical exponents. We give a compact review that first order RG in the couplings is sufficient to derive the exact transition from $\nu = 1$ to $\nu = 2/3$, if the boson radius obtained in the bosonization procedure is replaced by the exact radius obtained in the Bethe approach. We explain the fact, that from the bosonization procedure alone, the critical exponent can not be derived correctly in the isotropic limit $J_z \to J$. We further state that this fact is important if we consider to bosonize the antiferromagnetic super spin chain for the quantum Hall effect.

Keywords: Heisenberg spin 1/2 chain; spinchain; abelian bosonization.

PACS numbers: 11.10.Hi, 11.10.Kk, 73.43.-f

1. Introduction

This is a rather brief overview and the interested reader may consult more detailed literature, for instance the book of Giamarchi[1]. In order to calculate the critical exponent of the correlation length ν we generate a gap by introducing a dimerization term with coupling $\propto \delta$ and study the decay of the correlation length in the transition $\delta \to 0$. The dimerized spin 1/2 chain is then defined by the Hamiltonian:

$$H = -J \sum_i (1 - (-1)^{-i}\delta)[S_i^x S_{i+1}^x + S_i^y S_{i+1}^y + \frac{J_z}{J} S_i^z S_{i+1}^z]. \quad (1)$$

In the free limit $J_z \to 0$ the Hamiltonian is the dimerized xx-model and can be exactly diagonalized with the critical exponent of the correlation length $\nu = 1$. The Ising model is obtained in the limit $\delta \to 0$ and $J \to 0$ and the Heisenberg model (xxx-model) for $J_z \to J$ and $\delta \to 0$. In the Bethe ansatz[2,3,4] the critical exponent of the Heisenberg model is exactly known to be 2/3 and with the bosonization method it can be recovered, if the boson radius K is replaced by the exact radius obtained from the Bethe solution. Although the bosonization gives the exact critical exponent and RG-flow for the xx-limit, $J_z \to 0$, it fails to give the complete transition to the Heisenberg model without the exact solution.

2. Bosonization around the xx-Model

We briefly sketch the bosonization procedure around the xx-model. The Hamiltonian (1) can be mapped onto a half-filled chain of spinless fermions with the Jordan Wigner transformation. In one dimension this preserves all degrees of freedom of the spin chain and is therefore an exact transformation:

$$S_i^+ = (-1)^i e^{+i\phi_i} c_i^+ \quad \text{and} \quad S_i^z = c_i^+ c_i - \frac{1}{2}$$

$$\phi_i := \pi \sum_{n=1}^{i-1} c_n^+ c_n, \quad \text{where} \quad \langle e^{+/-i\phi_i} \rangle = (-1)^{N_F^i}.$$

N_F^i is the number of electrons at site 'i' and c_i are the fermion operators acting on the fermion vacuum $|0\rangle_i$ and satisfying the usual anticommutator relation $\{c_i, c_j^+\} = i/\hbar \delta_{i,j}$. In this scheme spin up is mapped to an occupied site, whereas spin down is mapped to an empty site. With the Baker Campbell Hausdorff formulas and operator identities the Hamiltonian (1) is transformed to the Hamiltonian

$$H = -\frac{J}{2} \sum_i (1 - (-1)^{-i}\delta)\left[c_i^+ c_{i+1} + c_{i+1}^+ c_i + \tfrac{2J_z}{J}(c_i^+ c_i - \tfrac{1}{2})(c_{i+1}^+ c_{i+1} - \tfrac{1}{2}) \right], \quad (2)$$

of spinless fermions on a half filled chain. We introduce the Fourier transformed operators $\{c_k, c_q^+\} = \delta_{k,q}$ on a finite lattice of length L with periodic boundary condition and lattice space a. Then, in the limit $J_z \to 0$ and $\delta \to 0$ the dispersion $\epsilon(k)$ of the free Hamiltonian

$$H = -\frac{2J}{a} \sum_k \cos(ka) c_k^+ c_k \qquad (3)$$

is linearized around the two Fermi points,

$$\epsilon(k) = \frac{2J}{a} \cos(ka) \quad \mapsto \quad \epsilon(k)_{R/L} = \pm \frac{v_F}{a}(k \mp k_F), \qquad (4)$$

and is proportional to the velocity $v_F = \partial_k \epsilon(k)|_{k_F}$ where R/L denotes left and right moving modes. In the linearized limit the densities

$$\rho(q) := \sum_k c_{k+q}^+ c_k \qquad (5)$$

satisfy then the bosonic current algebra:

$$[\rho(q), \rho(q')^+] = [\rho(q), \rho(-q')] = -\frac{Lq}{2\pi} \delta_{q,-q'}. \qquad (6)$$

To obtain the canonical commutation relation for bosons we introduce normalized bosonic operators:

$$b_q := \sqrt{\frac{2\pi}{Lq}} \rho(-q) \quad \text{and} \quad b_q^+ := \sqrt{\frac{2\pi}{Lq}} \rho(q) \quad \text{with} \quad [b_q, b_{q'}^+] = \delta_{q,q'}$$

We want to map the fermionic Fock space to a complete Fock space in terms of bosons, therefore we have to further introduce operators to connect Hilbert spaces with different numbers of particles. These so called Klein factors F commute with the boson operators:

$$F^+|N\rangle_0 = |N+1\rangle_0 \text{ and } F|N\rangle_0 = |N-1\rangle_0$$
$$[F, b_q^+] = [F, b_q] = [F^+, b_q] = [F^+, b_q^+] = 0.$$

In the case of different species of bosons i.e. left and right moving modes the Klein factors satisfy canonical anti commutation relations $\{F_\nu, F_\mu^+\} = \delta_{\mu,\nu}$, with $\nu, \mu = R, L$. However, we are interested in the thermodynamic limit, there the Klein factors can be mapped onto Majorana fermions $\{\eta_\mu, \eta_\nu\} = 2\delta_{\mu,\nu}$. In the continuum limit ($a \to 0$ and $L \to \infty$) the fermion operators c_i are replaced by their continuum counterpart $\psi(x)$, which can be expressed in terms of the bosonic operators b_q and the Klein factors with the so called Mattis-Mandelstam formula:

$$\psi_{R/L}(x) = \frac{F_{R/L}}{\sqrt{a}} e^{i\frac{2\pi N_{R/L}}{L}x} e^{\sum_{q>0} \alpha_q(x) b_{R/L,q}^+} e^{\sum_{q>0} \alpha_q^*(x) b_{R/L,q}} \qquad (7)$$

The coefficient $\alpha_q(x)$ is determined by the commutation relation

$$[b_{R/L,q}, \psi_{R/L}(x)] = -\sqrt{\frac{2\pi}{Lq}} e^{-iqx} \psi_{R/L}(x)$$
$$= \alpha_q(x) \psi_{R/L}(x) \qquad (8)$$

and the density is recovered by:

$$\psi(x) = \psi_R(x) + \psi_L(x)$$
$$\rho(x) = \rho_R(x) + \rho_L(x) + \psi_R^+(x)\psi_L(x) + \psi_L^+(x)\psi_R(x) \qquad (9)$$

To complete the bosonization procedure we introduce current fields $\theta(x)$ and charge fields $\phi(x)$:

$$\partial_x \phi(x) = \frac{1}{\sqrt{2}} \partial_x(\phi_R(x) + \phi_L(x)) = \frac{-\pi}{\sqrt{2}} \partial_x(\rho_R(x) + \rho_L(x))$$
$$\partial_x \theta(x) = \frac{1}{\sqrt{2}} \partial_x(\phi_R(x) - \phi_L(x)) = \frac{\pi}{\sqrt{2}} \partial_x(\rho_R(x) - \rho_L(x)). \qquad (10)$$

They satisfy the commutation relation in the continuum limit

$$[\phi, \tfrac{1}{\pi}\partial_y \theta(y)] = i\delta(x-y) \qquad (11)$$

and therefore the conjugate momentum operator is given by $\Pi(x) := \frac{1}{\pi}\partial_y \theta(y)$. We can now express the fermionic operator (7) in terms of the new operators:

$$\psi_{R/L}(x) = F_{R/L} \frac{1}{\sqrt{2\pi a}} e^{i(\pm k_F - \frac{\pi}{L})x} e^{-i(\pm \phi(x) - \theta(x))} \qquad (12)$$

The Hamiltonian then becomes in the continuum limit:

$$H_0 = \frac{v_F}{2} \int dx \left[\sum_\nu :(\nabla \phi_\nu)^2: + \frac{\pi v_F}{L} \sum_\nu N_\nu \right] \quad (13)$$

$$H_a = \frac{J_z/J}{L} \int dx \left[\sum_\nu N_\nu + 4 N_R N_L \right] \quad (14)$$

$$H_b = \frac{J_z/J}{\pi} \int dx \left[\tfrac{1}{2} \sum_\nu :(\nabla \phi_\nu)^2: + :\partial_x \phi_R(x) \partial_x \phi_L(x): \right] \quad (15)$$

$$H_c = -\frac{\delta(1+\frac{J_z}{\pi J})}{\pi a} \int dx \left[i F_L^+ F_R e^{i2\phi(x)} e^{i2\pi x N/L} + H.c. \right] \quad (16)$$

$$H_d = +\frac{J_z/J}{4\pi a^2} \int dx \left[F_R^+ F_R^+ F_L F_L e^{-i4\phi(x)} e^{i4\pi x N/L} + H.c. \right] \quad (17)$$

The first term in H_b renormalizes the Fermi velocity v_F and the Hamiltonian H_b plus the first part of H_0 together can be Bogoliubov transformed. The latter leads to the Hamiltonian:

$$H_K = \frac{\pi v_F}{2L}(1+g_4) \int dx \left[(1+\tfrac{g_2}{1+g_4}) N^2 + (1-\tfrac{g_2}{1+g_4}) J^2 \right] \quad (18)$$

where we introduced the so-called g-ology notation:

$$g_2 = 2 g_4 = \frac{2 J_z/J}{\pi v_F} \quad (19)$$

by rescaling the operators $N = N_R + N_L$, $J = N_R - N_L$ and introducing the boson radius K and the renormalized velocity u we obtain the standard Hamiltonian:

$$H_K = \frac{u}{2\pi} \int dx \left[K :(\nabla \theta(x))^2: + \frac{1}{K} :(\nabla \phi(x))^2: \right] \quad (20)$$

The boson radius in terms of the couplings becomes:

$$K = \sqrt{\frac{1+g_4-g_2}{1+g_4+g_2}} = \sqrt{\frac{1-\frac{1}{\pi}\frac{J_z}{J}}{1+\frac{3}{\pi}\frac{J_z}{J}}} \quad (21)$$

and the renormalized velocity u is given by:

$$u = \sqrt{(1+g_4)^2 - g_2^2} = \sqrt{(1+\tfrac{1}{\pi}\tfrac{J_z}{J})^2 - (\tfrac{2}{\pi}\tfrac{J_z}{J})^2} \quad (22)$$

In comparison, the exact boson radius and exact velocity obtained in the Bethe approach[2,3,4] are given by:

$$K = [2 - \tfrac{2}{\pi} \arccos(J_z/J)]^{-1} \quad (23)$$

$$u = \frac{\pi}{2} [1 - (J_z/J)^2] [\arccos(J_z/J)]^{-1} \quad (24)$$

and correspond for small values of $0 < J_z/J \ll 1$ to the bosonized values:

$$[2 - \tfrac{2}{\pi} \arccos(J_z/J)]^{-1} \approx \sqrt{\tfrac{1-\frac{1}{\pi}\frac{J_z}{J}}{1+\frac{3}{\pi}\frac{J_z}{J}}}\bigg|_{0<J_z/J\ll 1} \quad (25)$$

$$\approx 1 - \tfrac{2 J_z}{\pi J}\bigg|_{0<J_z/J\ll 1},$$

the velocity u is approximated similarly. In the thermodynamic limit the Klein-factors can be replaced by Majorana fermions and we rewrite H_c and H_b accordingly:

$$H_c + H_d = -\frac{2\delta(1+\frac{J_z}{\pi J})}{\pi a}\int dx\ \cos(2\phi(x)) + \frac{J_z/J}{2\pi a^2}\int dx\ \cos(4\phi(x)) \quad (26)$$

The first Hamiltonian H_c is the dominant part from the dimerization and the second is the marginal, so called umklapp term. In the thermodynamic limit the complete Hamiltonian has become a sine-Gordon model:

$$H = H_K + H_c + H_d. \quad (27)$$

3. Perturbative RG and the Exact Critical Exponent

The sine-Gordon model treated in the perturbative renormalization group approach gives the exact transition of the critical exponent, from the free model to the Heisenberg model, as long as we replace the boson radius by the exact radius from the Bethe solution. We start with the general model

$$S = \int d^2x\ \left[\tfrac{1}{2}(\nabla\phi(x))^2 - \sum_i \lambda_i \cos(\beta_i \phi(x))\right], \quad (28)$$

where we introduced a normalized gradient $\nabla = (\tilde{\partial}_0, \tilde{\partial}_1)$ with the derivatives $\tilde{\partial}_i := \sqrt{m_\mu}\partial_\mu$ and the masses m_μ define the boson radius $K := 1/\sqrt{m_0 m_1}$. In order to obtain the RG equations for the coupling constants, we have to integrate out the fast moving modes (UV modes). Therefore we split the field $\phi(x)$ into an IR part $\phi'(x)$ and an UV part $\delta\phi(x)$:

$$\phi(x) = \phi'(x) + \delta\phi(x), \quad \phi(x) = \begin{cases} \phi(x), & 0 < k \le \Lambda' \\ \delta\phi(x), & \Lambda' < k \le \Lambda. \end{cases} \quad (29)$$

With this definition the partition function can be separated into two traces:

$$Z_\phi = \mathrm{Tr}_\phi[e^{-\mathcal{H}}] = \mathrm{Tr}_{\phi'} \circ \mathrm{Tr}_{\delta\phi} e^{-\mathcal{H}} \quad (30)$$

and the normalized trace of the fast moving modes is given by:

$$\langle \cdot \rangle_{\delta\phi} = Z_{0,\delta\phi}^{-1}\ \mathrm{Tr}_{\delta\phi}[e^{-\int d^2x\ \frac{1}{2}(\nabla\delta\phi(x))^2}(\cdot)] \quad (31)$$

Now freeze out the fast moving modes by calculating:

$$\langle e^{-\mathcal{H}}\rangle_{\delta\phi} = \exp\left\{-\int d^2x\ \tfrac{1}{2}(\nabla\phi'(x))^2 - \sum_i \lambda_i\ e^{(2-\frac{\beta_i^2 K}{4})l}\cos(\beta_i\phi'(x))\right\}. \quad (32)$$

We introduced the length $l := \ln(\Lambda'/\Lambda)$ and rescaled the integration measure $dx_0 dx_1 \mapsto e^{2l}dx_0 dx_1$. The first order renormalization group equations are obtained as the variation of the coupling constants with respect to the length l:

$$\frac{d\lambda'_i}{dl} = \left[2 - \tfrac{1}{4}\beta_i^2 K\right]\lambda'_i \quad (33)$$

For the spin chain the critical exponent is directly determined by these equations. For small J_z the RG equation of the coupling λ'_1 gives the critical exponent, with

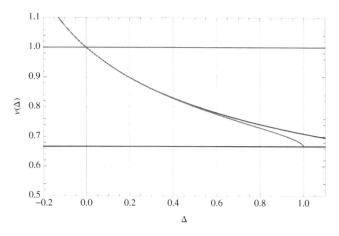

Fig. 1. The transition of the critical exponent ν in terms of the coupling parameter $\Delta = J_z/J$. The red curve is produced by the boson radius K from the bosonization, while the brown curve is the exact result from the Bethe solution. To larger $J_z \to J$ we obtain a deviation from the exact result. Plotted with Mathematica 7.

corresponding $\beta_1 = 2$ and K from above:

$$\frac{d\lambda'_1}{dl} = \left[2 - \sqrt{\frac{1 - \frac{1}{\pi}\frac{J_z}{J}}{1 + \frac{3}{\pi}\frac{J_z}{J}}}\right]\lambda'_1. \tag{34}$$

In the limit $J_z \to 0$ this gives the exact solution of the critical exponent $\nu = 1$ with the correct flow for small J_z/J. However, in the isotropic or $SU(2)$ limit it fails to produce the critical exponent of $\nu = 2/3$ it gives rather more like $\nu \approx 0.709$, see Fig. 1. This can be fixed if we take the exact results for the boson radius K from the Bethe ansatz (23) for the Heisenberg spin chain. If we replace the boson radius from the bosonization with the exact expression, we obtain a continuous transition of the critical exponent form $\nu = 1$ in the xx-limit to $\nu = 2/3$ in the xxx-limit:

$$\xi \propto \delta^{-\nu}, \ \nu = \left[2 - \frac{\pi}{2(\pi - \arccos(J_z/J))}\right]^{-1}. \tag{35}$$

This transition can also be recovered, if the exact boson radius is used in the self-consistent harmonic analysis instead of the first order RG-approach[5]. If we go to second order RG, then the boson radius itself becomes dynamic:

$$\frac{dK^{-1}}{dl} = \frac{1}{4}\sum_{i,j}\lambda_i\lambda_j\beta_i\beta_j\ K\ C_{ij}(K,\Lambda), \tag{36}$$

with the integral C_{ij} depending on K and the choice of the cutoff Λ, further details can be found in the book of Giamarchi[1]. In the spin chain the coupling J_z couples intrinsically the couplings K and λ_2. In second order both become dynamic and the flow of the boson radius depends on the flow of both couplings λ_i. However, we can not expect to recover the exact result of K in the limit $J_z \to J$. This is the regime where perturbative methods usually break down as long as we can not sum up[...] fact is reflected in the exact K. Since its derivative diverges in t[...] the perturbative RG approach has to be performed up to all orders[...] cond order RG it is not possible to recover the exact solution for the coup[...]

4. Outlook

We reviewed the RG approach for the antiferromagnetic spin 1/2 chain and pointed out that the exact critical exponent and RG-flow for small coupling $0 < J_z/J \ll 1$ can be derived within the abelian bosonization approach and to first order RG, while in the strong limit it rather gives a rough upper limit. However, the exact transition of the critical exponent to the isotropic $SU(2)$ symmetric Heisenberg model is obtained, if the correct boson radius is known from the Bethe solution. In the antiferromagnetic super spin chain for the quantum Hall effect, the pure fermionic sector can be derived via the Chalker Coddington network model[8] or via the Landauer conductance[7] and leads to the antiferromagnetic spin 1/2 Heisenberg chain. If we (abelian) bosonize the fermionic sector we have also to replace the perturbative boson radius by the exact one in the strong coupling limit. Furthermore, since we mapped the fermions onto compact bosons, we might consider to compactify the bosonic sector as well. Recently a compact theory with twisted fermions[9] has been suggested for the integer quantum Hall transition, which gained support from numerical studies[10].

Acknowledgments

Grateful acknowledgment for financial support goes to the research center MAMOC at Jacobs University Bremen and the WCU project in the division of Advanced Materials Science at POSTECH University in Pohang, South-Korea.

References

1. G. Thierry. Quantum Physics in One Dimension *Oxford University Press* 2003.
2. R. J. Baxter. Eight-vertex model in lattice statistics. *Phys. Rev. Lett.*, 26(14):832–833, 1971.
3. M. C. Cross and D. S. Fisher. A new theory of the spin-peierls transition with special relevance to the experiments on ttfcubdt. *Phys. Rev. B*, 19(1):402–419, 1979.
4. A. Luther and I. Peschel. Calculation of critical exponents in two dimensions from quantum field theory in one dimension. *Physical Review B*, 12(9), 11 1975.
5. T. Nakano and H. Fukuyama. Dimerization and solitons in one-dimensional xy-z antiferromagnets. *Journal of the Physical Society of Japan*, 50(8):2489–2499, 1981.
6. K. Nomura. Logarithmic corrections of the one-dimensional S=1/2 Heisenberg antiferromagnet. *Phys. Rev. B*, 48:16814–16817, 1993.
7. A. Sedrakyan. Action formulation of the network model of plateau-plateau transitions in the quantum hall effect. *Phys. Rev. B*, 68:235329, 2003.

8. M. R. Zirnbauer. Conformal field theory of the integer quantu[...]tion. 1999.
9. C. A. Lütken and G. G. Ross, Geometric scaling in the quantum H[...]ics Letters B, 653, 2-4, 363–365, Sep 2007.
10. S. Keith and O. Tomi, Critical exponent for the quantum Hall transi[...]ys. Rev. B 80:041304, 2009.